OVER 400 SPECIES IN FULL COLOR

NON-FLOWERING PLANTS

by

FLOYD S. SHUTTLEWORTH

and

HERBERT S. ZIM

ILLUSTRATED BY

DOROTHEA BARLOWE,

SY BARLOWE, JACK KUNZ, BARBARA WOLFF,

AND JEAN ZALLINGER

A GOLDEN NATURE GUIDE

GOLDEN PRESS • NEW YORK

Western Publishing Company, Inc.

Racine, Wisconsin

FOREWORD

The world of non-flowering plants includes some of the most fascinating, yet least understood of all plants. Many bacteria cause plant and animal diseases, yet other bacteria and some fungi bring about the decay of plant and animal remains, thereby keeping their chemicals in cycle. Great conifer forests encircle the earth in the cooler regions and are a measure of natural resource in the economy of many nations. Seemingly insignificant lichens are pioneer plants that convert weathered rocks into soil. Many species of mushrooms are delicacies prized by gourmets; others are deadly poisonous. Thus, non-flowering plants are both interesting and important. This book contains only a selection of the many thousands of species.

Special thanks go to the artists, particularly to Dorothea Barlowe, who has drawn the majority of the species; to Sy Barlowe, Barbara Wolff, and Jean Zallinger, each of whom has made a number of the illustrations, and to Jack Kunz, who did the cover art. The authors are also gratefully indebted to Dr. William A. Weber, Professor of Natural History, University of Colorado Museum, to Dr. William C. Steere, Director of the New York Botanical Garden, and his associate, Dr. Clark T. Rogerson, to Dr. Taylor R. Alexander, Mrs. Lillian Fly, Mrs. Julia Morton, and Dr. Joseph M. Riedhart, of the Department of Botany, University of Miami, and to the many others who made helpful suggestions and furnished various materials.

F.S.S.; H.S.Z.

 Printed in the U.S.A. Published by Golden Press, New York, N.Y. Library of Congress Catalog Card Number AC 67-16476.

CONTENTS

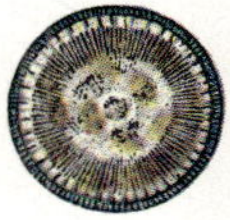

INTRODUCING NON-FLOWERING PLANTS

No single coherent group among the more than 350,000 species of plants can be labeled as non-flowering plants. Yet about one third—over 100,000 species—reproduce without bearing flowers. These range from microscopic bacteria to the largest of all trees, the redwoods. Included are the various groups of algae and fungi, mosses, liverworts, hornworts, ferns and their allies, cycads, conifers, and scores of smaller, rarer groups.

Botanists classify plants in an evolutionary sequence by placing those with similar structures and life histories together, as far as present knowledge permits. This task is not easy, for plants that superficially appear alike may have developed from different ancestors. Structures that have changed gradually over millions of years may be difficult to trace because of the scanty fossil record.

This is especially true of the earliest living things. Organic materials occur in rocks well over two billion years old. The earliest life may have resembled present-day viruses, which exhibit only the basic characteristics of life. Viruses either are or contain complex nuclear proteins, RNA or DNA, and reproduce in living plant or animal cells. Some viruses are of a crystalline nature and may be only self-duplicating molecules. Others are more clearly "living."

No link between bacteria and viruses has been established. Bacteria seem too advanced to have been the earliest living organisms. But they can subsist on simple organic compounds of the sort that may have been formed by the action of lightning on water and early atmospheric gases.

Other early evolutionary steps are equally hazy. Perhaps bacteria, blue-green algae, and flagellates (single-celled organisms that swim with a whiplike appendage) did develop from something like a primitive virus. The other algal classes may have arisen separately from the flagellates, or the red algae may have come from the blue-greens and others from the greens. Fungi may have come directly from the flagellates. Bryophytes and later the vascular plants (those with a conducting system) may have developed directly from green algae. Hornworts, though rootless and without true leaves, developed conductive tissues. True vascular tissues appeared in early *Psilotum*-like plants. Over millions of years, these plants became adapted to growth on land and gave rise to more advanced non-flowering plants—the ferns, fern allies, and gymnosperms—and, finally, to the flowering plants.

LAND PLANTS with supporting and conducting tissues (vascular) first appeared about 400 million years ago. Below are fossils of common, non-flowering kinds, some of them found with coal.

Seed Fern

Lycopod

Horsetail

Ginkgo

Generally, plant evolution has resulted in larger and more complex organisms. In the first aquatic plants, tissues for support and conduction were not necessary. On land, fundamental changes were needed to provide support and to transport food and water throughout the plant body. Then with the development of pollen and seeds in gymnosperms, the dependence of plants on water for fertilization was eliminated, and they became more widely distributed on land.

THE RELATIONSHIPS among groups of non-flowering plants, reflecting their evolution, are not yet fully known. The family tree on these pages represents a fusion of several points of view. In this book, the treatment of the various plant groups (p. 3) does not follow the detailed sequence suggested on these two pages. Instead, the euglenoids, dinoflagellates, stoneworts, and diatoms have been placed with the other algae—blue-green, green, red, and brown. This brings together groups that possess chlorophyll and make their own foods. In the same way bacteria and slime molds have been treated with the fungi. Generally, all lack chlorophyll and obtain food from living plants and animals as parasites, or from dead matter as saprophytes.

THALLOPHYTES

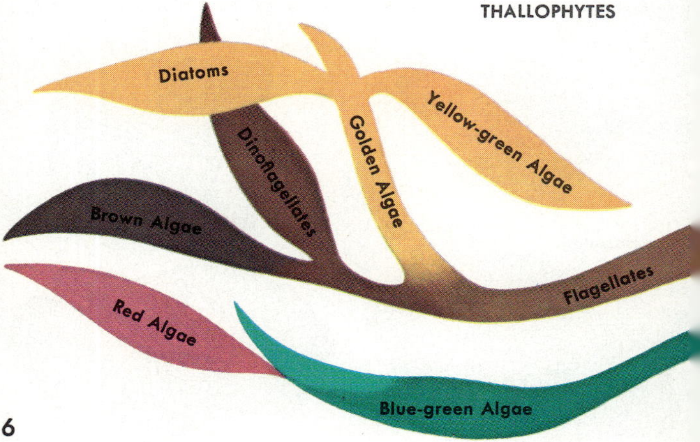

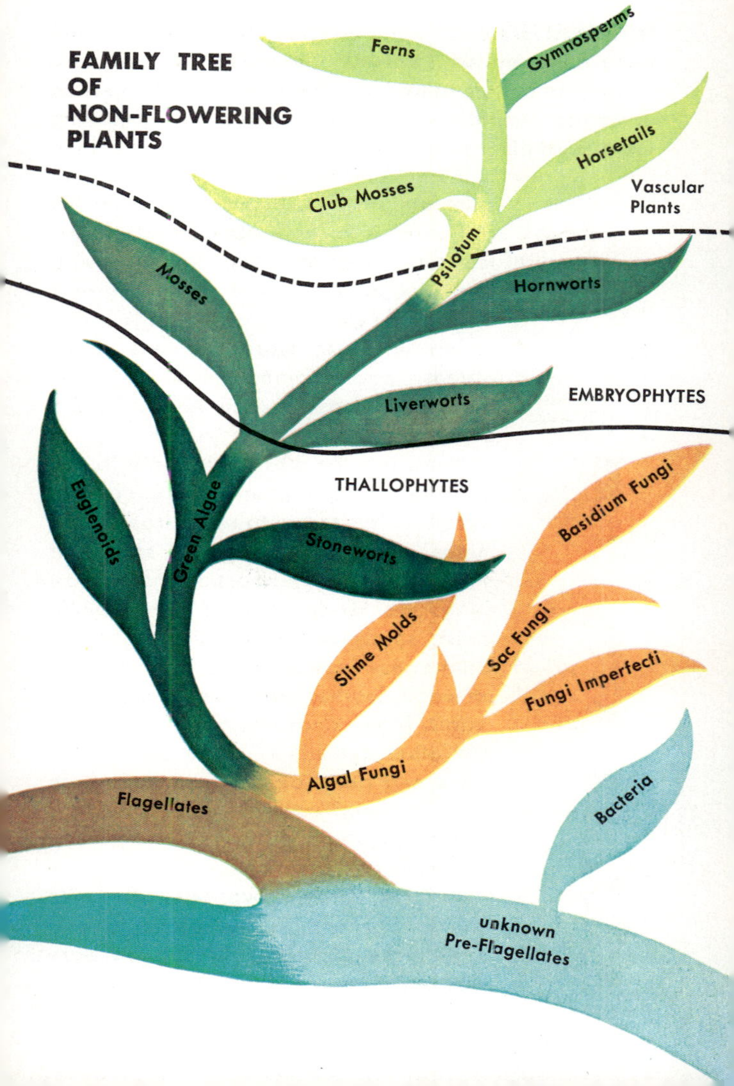
FAMILY TREE
OF
NON-FLOWERING
PLANTS
Ferns
Gymnosperms
Horsetails
Club Mosses
Vascular
Plants
Psilotum
Mosses
Hornworts
Liverworts
EMBRYOPHYTES
THALLOPHYTES
Euglenoids
Green Algae
Stoneworts
Basidium Fungi
Sac Fungi
Slime Molds
Fungi Imperfecti
Algal Fungi
Flagellates
Bacteria
unknown
Pre-Flagellates

DISTRIBUTION of non-flowering plants varies with geography and with habitat. Generally they are more widespread than flowering plants and less subject to ecological limitations. Moisture is often the controlling factor in their distribution.

Red Alga

Yellow-green Alga

Mushroom

Lichen

Algae are most common in fresh or salt water, though some may grow on wet soil, bark, or rocks. Their distribution varies from group to group. Reds, browns, and dinoflagellates are mainly marine. Yellow-green and golden algae live mainly in fresh water. Greens, blue-greens, and diatoms grow in both, though more species occur in streams and ponds.

Fungi are the most widely distributed of non-flowering plants. They do well from cool temperate to warm tropical climates. Bacteria, yeasts, and spores of molds are in the air about us and on most things we touch. Many fungi grow in moist, shady soil; some live in water. Others grow in manure and leaf mold and on wood, leather, and other organic materials. Many kinds are parasitic on man and other animals or cause diseases of crops and other plants. Together, these parasitic fungi cause millions of dollars in damage yearly.

Lichens also have a very wide distribution. Some species cover mile after mile of open ground in the arctic tundra. Lichens are pioneers, thriving on bare rock, sand, or soil. They resist drying.

Mosses are usually shade-loving plants of open woods, fields, and river banks. Some species grow in sand, in rock crevices, and on fallen timber. Mosses are common in temperate climates and at higher altitudes in the tropics. A few grow in water.

Liverworts and hornworts thrive in moist soil, forming green blankets along streams and in other wet places. These lesser plants are often studied as examples of alternation of generations.

Ferns are more widespread and more abundant in the wet tropics than in more temperate areas. Most species prefer moist, shaded habitats, though some can stand the open sun of fields and roadsides. North America has a number of bog species and a few kinds that grow in ponds and streams. Some ferns of warmer areas are epiphytes.

Fern allies, closely related to ferns, grow in open or shaded woods; some groups thrive along roadsides, on embankments, and even on sterile, filled land. A few are also aquatic. One in semiarid areas curls up to resist drying.

Gymnosperms, the most advanced of the non-flowering plants, thrive in diverse environments. Best known are the great boreal forests of conifers, but members of this group also form forests in Australia and South America. Cycads and several lesser known groups are most abundantly represented in the tropics and subtropics.

Moss

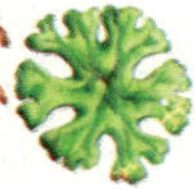

Liverwort

Fern

Club Moss

Cycad

ACTIVITIES

Non-flowering plants are best studied where they are growing naturally. Make notes of the species you find and where you find them. Your records will become increasingly valuable as they are extended with each trip. People interested mainly in identifying non-flowering plants often collect specimens. They find that photographs, taken in the field or indoors with controlled lighting, are helpful in identifying plants and studying growth habits. Others may enjoy growing plants in the simulated natural environment of a terrarium. Those gastronomically inclined may be attracted to wild mushrooms. A number of seaweeds are also palatable. Experiments on use of such green algae as *Chlorella* for food may help solve the problem of feeding space travelers.

SPORE PRINTS provide a permanent record of spore color and gill pattern of larger fungi. The preserved prints not only make an interesting collection but are valuable in identifying difficult species. Follow the technique illustrated below. Try various colored papers to get the best contrast between spores (many are white) and the background. Spray the finished print with lacquer or a fixative to prevent smearing. Label each.

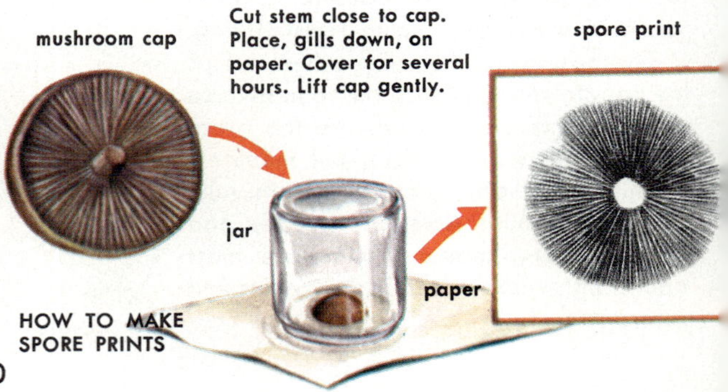

HOW TO MAKE SPORE PRINTS

PHOTOGRAPHS provide an interesting, artistic, and effective record of non-flowering plants. Because dried plant specimens lose color and form, photographs of fresh specimens are an excellent aid in identification. Since most plants are small, you will need an extension tube or portrait lens. A telephoto lens is also valuable. Bring specimens indoors and photograph them against a plain background. Better still, photograph them where they grow. For this you will need a tripod, flash or strobe light, and plenty of patience. Plants rarely grow in photogenic positions, but with effort, you can get identifying details in your pictures. Consider adding a coin or a small ruler to show scale.

Improve your photography by keeping a record of settings and conditions for every shot. Check your print with your notes on exposure, light, and film. Follow through and correct your weak points the next time you take your camera afield.

Sea Lettuce photographed in a white pan for contrast.

Ferns and lichens are good plants for terraria.

TERRARIA may be used to grow small non-flowering plants. You can observe details and see fruiting bodies develop. Photography is easier, too, with specimens grown in these movable habitats. Terraria provide small, controlled environments that give plants growing conditions much like those of their natural habitats. To make a terrarium, select a fish bowl, jar, or other large glass container. Cover its bottom with damp sand and charcoal in alternate thin layers. Put in small plants like liverworts, lichens, fungi, tiny ferns, or fern allies. Moisten, but not too often, as the lid prevents evaporation. Be sure to keep the terrarium out of direct sunlight.

Closeup shows lower and upper surfaces of a gill fungus.

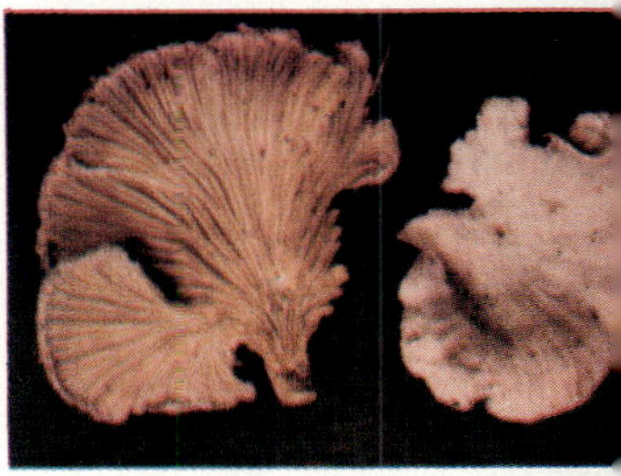

COLLECTIONS make it possible to study plants in more detail than can be done in the field. Specimens from different trips and areas can also be compared.

lichen packet

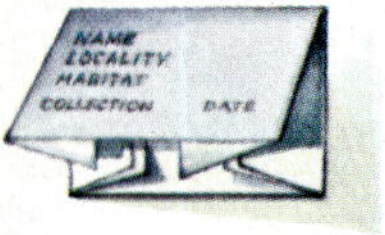

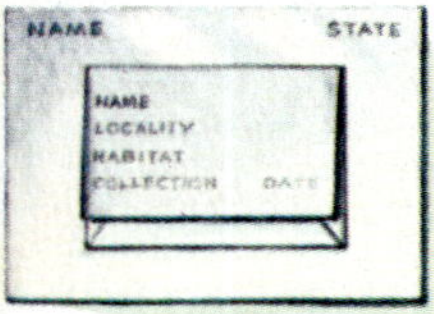

glued to card

filed in shoebox

SEAWEEDS and large fresh-water algae can be preserved in alcohol (either denatured or rubbing) or in 5 percent formaldehyde. Specimens can also be mounted on a heavy paper. First float the seaweed in a pan of water. Push the mounting paper beneath. Spread and arrange the specimen as you slowly raise the paper, lifting the specimen out. Cover each sheet with damp cheesecloth and dry between newspapers.

WOODY FUNGI specimens are nearly dry when collected, but fleshy fungi must be sun- or oven-dried to preserve them. Each dried specimen can be stored in a separate box, with a data card of pertinent information. To protect specimens from insect pests, put mothball or "para" crystals in box.

LICHENS can be air-dried or pressed between blotters. Each dried specimen is put in a 3x4-in., folded packet or in an envelope, with collection data written on flap. Glue envelope to card, label, and file.

MOSSES, LIVERWORTS, FERNS, AND FERN ALLIES are usually pressed, dried, and mounted on herbarium sheets. Small specimens may be kept in labeled packets.

collection data

fern mounted on herbarium sheet

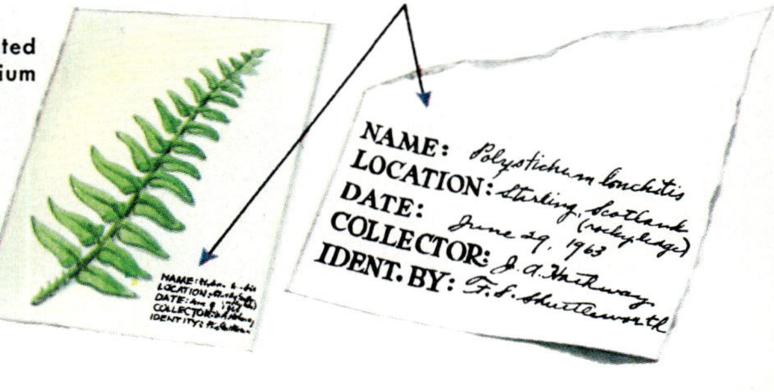

NECTRIA, a weakly parasitic sac fungus (p. 34), causes a disease of apple and other trees. The red fruiting bodies cover the thallus. Dia. 0.1-0.2 in.

THALLOPHYTES

The plant kingdom has been divided into the Thallophytes and the Embryophytes. In the former division are the various groups of algae, fungi, and lichens. The Embryophytes, a more complex group, include the remaining classes of non-flowering plants in addition to the flowering groups. Thallophytes total some 100,000 species, which fall into 15 or more classes. A general picture of the evolutionary relationships among the various Thallophyte groups is shown on pp. 6-7.

The plant body of a Thallophyte is a thallus, which may be a single cell or a larger mass of relatively undifferentiated cells that show little specialization except for those involved in reproduction. Thallophytes lack conductive (vascular) tissue. They have no roots, stems, or leaves; nor do they produce flowers, seeds, or fruits.

Thallophytes typically have single-celled reproductive structures. Those that are many-celled lack a jacket of sterile cells around sex cells and around the spores. The stoneworts are an exception. The fertilized egg (zygote) of Thallophytes does not develop into an embryo. Reproduction also occurs asexually by cell division or by formation of spores of many types, motile and non-motile.

ALGAE

Algae are Thallophytes that contain chlorophyll and are therefore able to manufacture their own food. They include 10 or more classes (see chart on opposite page), varying from microscopic one-celled organisms to kelps that grow to over 200 feet long.

Nearly all algae live in water. Many reproduce both asexually and sexually. Free-swimming asexual spores (zoospores) propelled by flagella are common, but many algae produce various kinds of non-motile asexual spores. In some algae, the sex cells look alike but are physiologically different. In others, eggs and sperms are produced in separate cells or organs.

Algae are directly or indirectly an important source of food for all aquatic animals. They help maintain the oxygen content of the water by releasing oxygen as a by-product of photosynthesis. Some algae are undesirable, as they contaminate water supplies and swimming pools. A few produce toxins poisonous to cattle that drink the water. Some of the marine forms are used as food. Red algae are the source of agar, important in the laboratory culture of bacteria. Alginic acid, used

REPRODUCTION IN ALGAE

ASEXUAL

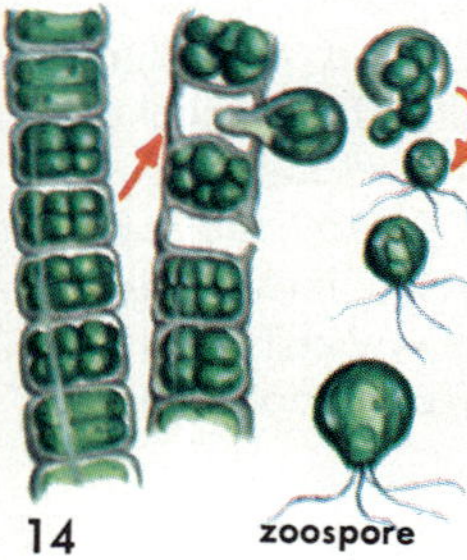

Asexual—parent cell becomes modified, ruptures, and releases zoospores, as in *Ulothrix* at left.

Sexual—some cells develop large eggs; others form sperm that swim to egg and fertilize it, forming a zygote—as in *Oedogonium* at right.

SEXUAL

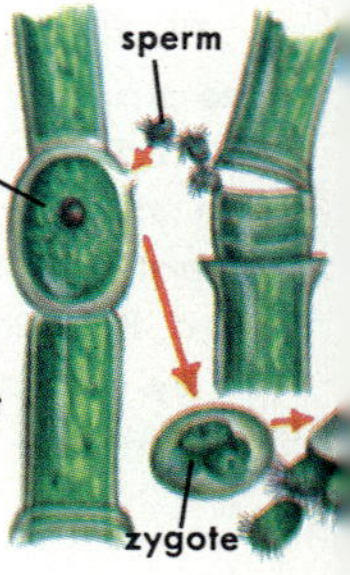

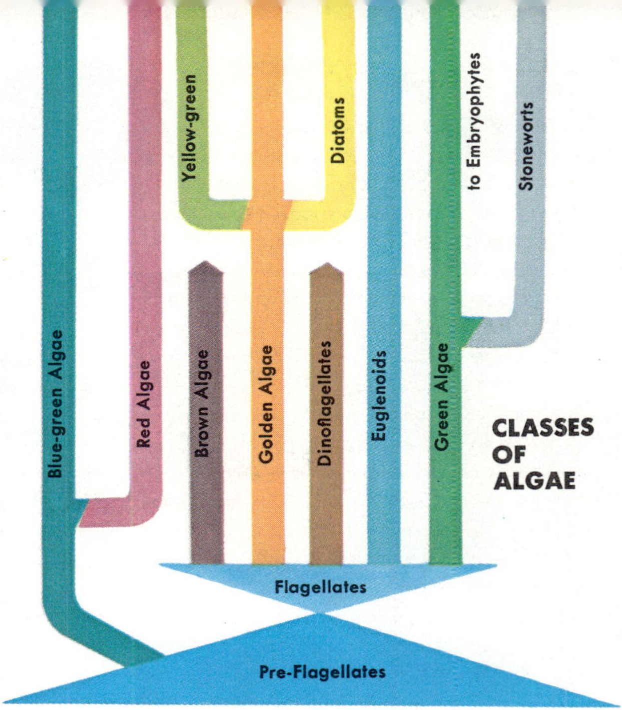

in ice cream, is obtained from brown algae. Diatomaceous earth has many uses (p. 20).

Marine algae, especially the larger red, brown, and green species, are treated in the Golden Nature Guide *Seashores* and in other books listed on p. 153. On the following pages are samples of some important genera of marine algae. Except for relatively few of the larger, conspicuous species, identification is difficult and may require microscopic study.

MARINE ALGAE

GREEN ALGAE

Caulerpa
2-3 in., grows
from long runners
in tropical waters

Merman's
Shaving Brush
6 in.,
tropical
waters

Sea Lettuce
2-16 in. sheets,
widespread

Mermaid's Wine Glass
2-4 in.,
tropical waters

BROWN ALGAE

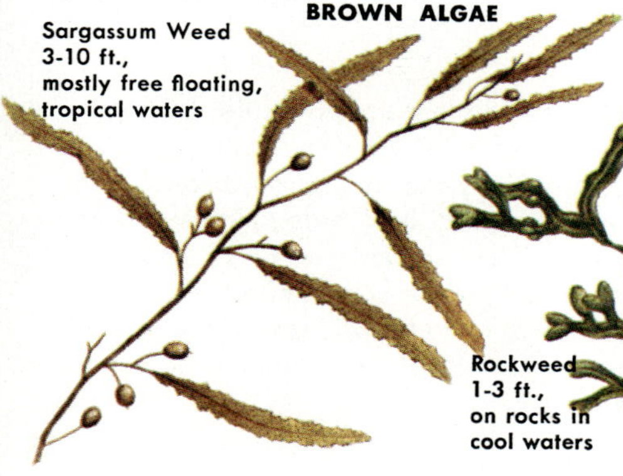

Sargassum Weed
3-10 ft.,
mostly free floating,
tropical waters

Rockweed
1-3 ft.,
on rocks in
cool waters

BROWN ALGAE

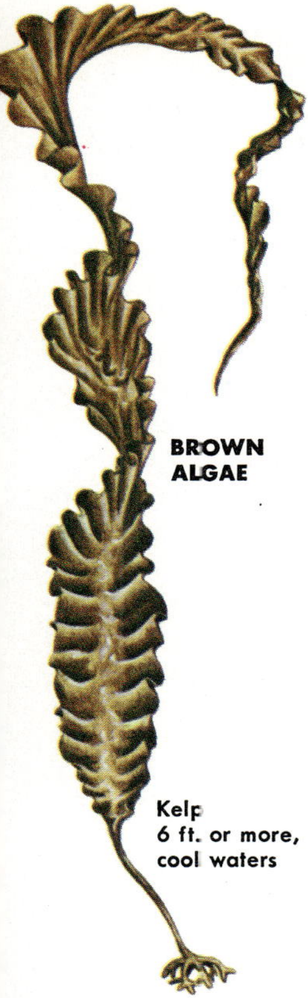

Kelp
6 ft. or more,
cool waters

Padina
2-4 in.,
tropical

RED ALGAE

Ceramium, 2-9 in.,
shallow tropical waters

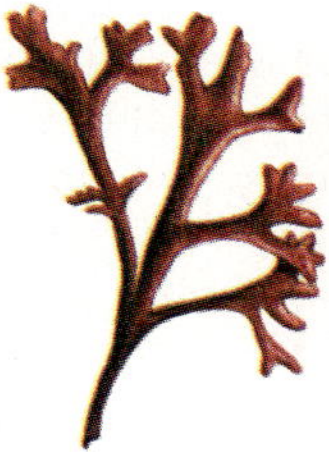

Irish Moss
clustered, 3 in.
cool waters

Dasya
8-20 in.,
widespread in
shallow waters

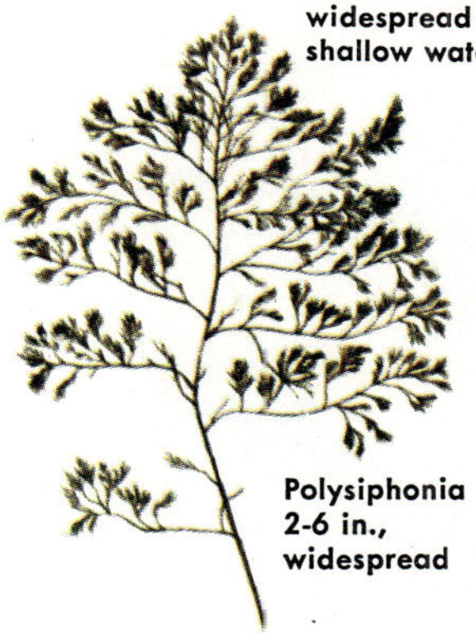

Polysiphonia
2-6 in.,
widespread

Mermaid's Hair (natural size) is common in shallow salt water, covering rocks and pilings.

BLUE-GREEN ALGAE (about 1,500 species) are widespread in salt and fresh water. Most are covered by a gelatinous sheath, enabling them to live in hot springs, in polluted water, or on moist tile and soil. In late summer some become so abundant they form a scum on ponds. Some poison water for drinking.

Blue-green algae are the simplest of all algae. They have a poorly defined nucleus. Color pigments are unconfined, spreading over the cells. Reproduction is by simple cell division. Some blue-greens occur as single microscopic cells; some in filaments; others as gelatinous masses. *Chroococcus* has distinct gelatinous sheaths. *Merismopedia* hangs together in flat colonies. *Oscillatoria,* with its disklike cells arranged in a filament, resembles a stack of coins. *Nostoc* has strings of beadlike cells, and *Gloeotrichia,* also filamentous, may be free-floating or attached. When abundant, *Gloeotrichia* can irritate the skin of swimmers.

highly magnified

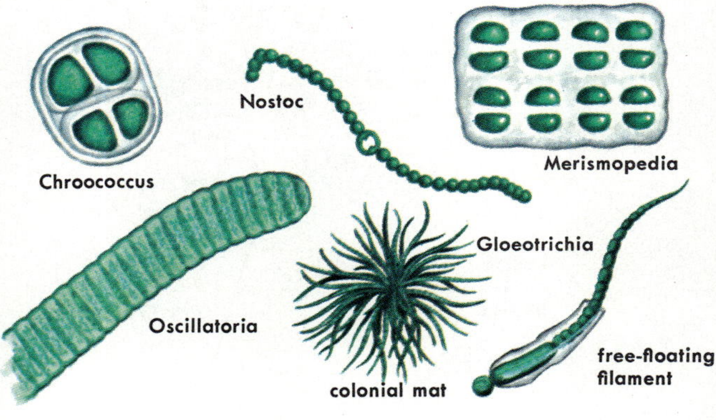

RED ALGAE (about 3,000 species) are mainly fairly small and delicate marine species (p. 17). A few occur in fresh water. All red algae contain chlorophyll, but because the green is masked by other pigments, most of these algae appear pink or purplish. Several species can be eaten.

Batrachospermum (natural size)

one of several species of fresh-water red algae

BROWN ALGAE (1,500 species) are common in cool marine waters (pp. 16-17). Some of the kelps yield algin, a chemical used as a thickening or smoothing agent in foods. Most are attached to rocks by holdfasts, but Sargassum Weed often floats in huge mats.

GOLDEN ALGAE (about 300 species) usually have one or two golden-brown chloroplasts (chlorophyll-containing bodies). The cells are surrounded by membranes or have walls containing a small amount of silica. Most are motile. A few species occur in salt water, but most are found in lakes, pools, and ditches.

SYNURA's naked cells are joined to form a spherical colony. Each cell has two flagella. Common in lakes and ponds.

DINOBRYON forms treelike colonies. Each of the conical cells is surrounded by a transparent sheath.

highly magnified

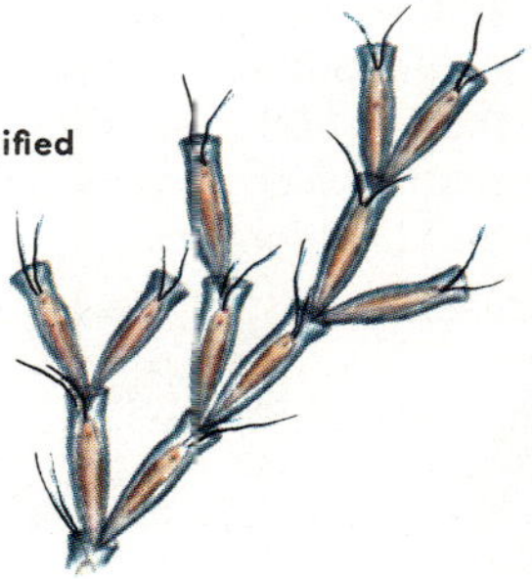

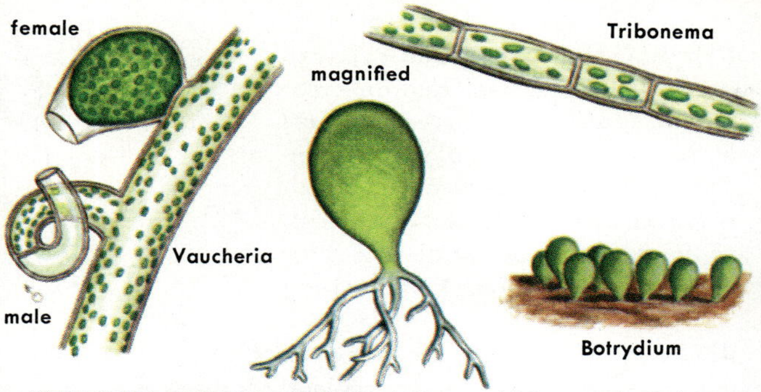

YELLOW-GREEN ALGAE (about 400 species) have yellow pigment and store reserve foods as oils rather than as starch. Most live in fresh water, a few on soil. *Vaucheria,* a tubular alga, is commonly aquatic, but some species form feltlike mats on wet soil. *Tribonema,* a filamentous alga of ponds and ditches, has cell walls of hard-to-see, overlapping, H-shaped pieces. *Botrydium,* about the size of a pinhead, grows on mud or on moist soils.

DIATOMS (perhaps 10,000 species) are single-celled algae with cell walls that overlap like a loosely covered box. The walls, often beautifully ornamented, are impregnated with silica. The silica does not disintegrate when the algae die, and so the "skeletons" accumulate on the bottoms of lakes and bays. Some deposits, more than 1,000 feet thick, are mined for diatomaceous earth, which is used in filters, for insulation, in polishes, and in cleansers.

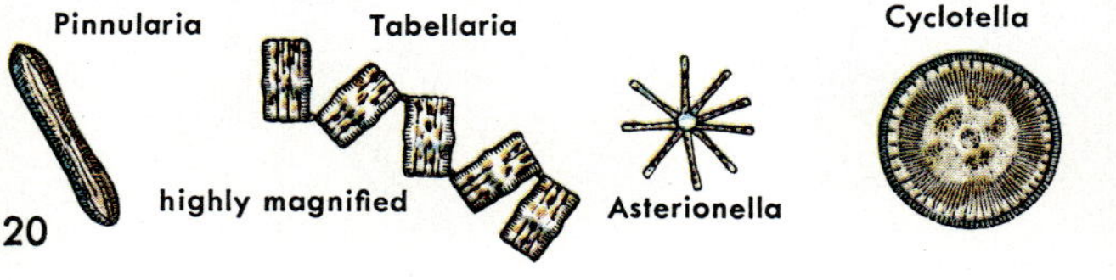

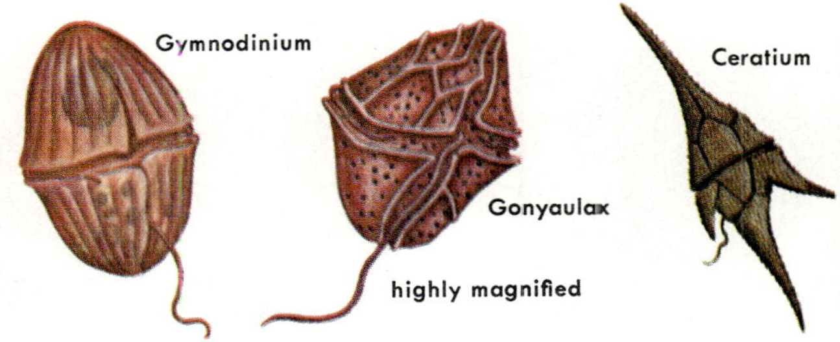

GYMNODINIUM causes poisonous "red tide." *Gonyaulax* makes shellfish poisonous. *Ceratium* is a fresh-water dinoflagellate.

DINOFLAGELLATES (about 1,000 species) have both plant and animal characteristics. Most dinoflagellates can swim. Some have only a cell membrane, but many have cell walls made of a series of plates with two deep grooves, one along the length of the cell and the other around it. Each has two flagella, a short one that lies in the horizontal groove and a longer one that trails along the vertical groove. Most dinoflagellates are marine; a few prefer fresh water.

EUGLENOIDS (about 400 species) are all single-celled and motile. The majority contain chlorophyll and thus can carry on photosynthesis. Unlike most other unicellular plants, however, they lack a cell wall and have other animal characteristics. *Euglena* sometimes occurs so abundantly that stagnant pools are colored deep green. It has a red eyespot and can change its shape. *Phacus,* with a fixed, somewhat flattened shape, is found in fresher water.

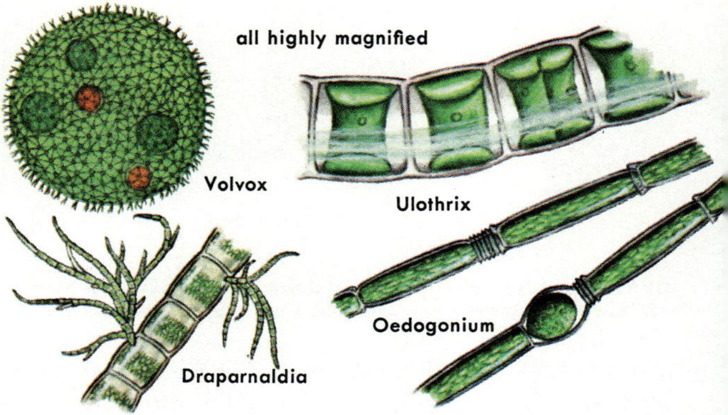

GREEN ALGAE (about 6,500 species) were probably the group from which land plants began to evolve. Their structures and life cycles show basic similarities to those of more advanced plants. Each green algal cell has at least one chloroplast—the site of photosynthesis. Food is stored as starch. Chlorophyll pigments just like those of higher plants are found in the variously shaped chloroplasts. Green algae may be single- or many-celled, and they exist on land as well as in salt and fresh water. Most are small, unnoticed except in floating masses or as clustered growths on rocks, but some marine forms may be a yard or more long. Both sexual and asexual reproduction common.

Fresh-water *Volvox,* a spherical colonial form with both plant and animal characteristics, may reach the size of a pinhead. It rolls through water, propelled by the whipping motion of flagella, two projecting from each of the colony's many cells. *Ulothrix, Oedogonium,* and *Draparnaldia* have cells arranged in hairlike filaments. The first two grow attached to sticks, stones, and plants. The many-branched *Draparnaldia* forms slippery masses in cool springs.

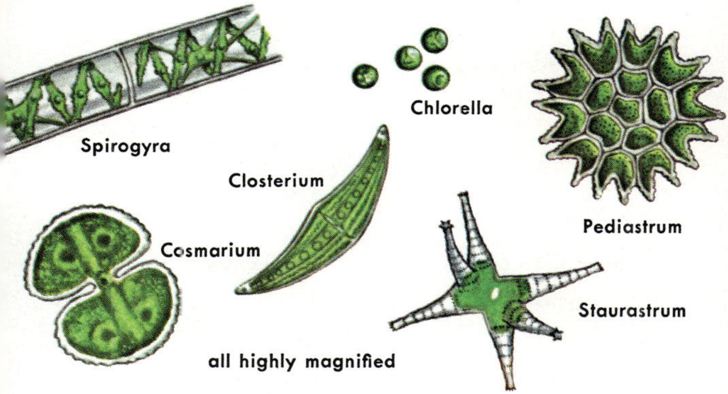

all highly magnified

Spirogyra, sometimes called Pond Silk, is a widespread green alga with ribbon-like, spiral chloroplasts. *Cosmarium, Closterium,* and *Staurastrum* are closely related to *Spirogyra,* but all are single cells with each half-cell having a distinctive chloroplast. The colonial *Pediastrum* may grow in ditches. Spherical cells of *Chlorella,* used in studies of respiration and photosynthesis, are common in fresh and salt water.

STONEWORTS (more than 200 species) are a small group of marine and fresh-water algae. Though sometimes classified with the green algae, they share with Embryophytes such features as a jacket of sterile cells around the sex cells. Stoneworts are anchored in the bottom mud by slender rhizoids (rootlike filaments). Reproductive structures can be seen as red bumps on the whorls of branches that arise from the slender, smooth green stems. Stoneworts that grow in limestone regions are usually encrusted with lime.

natural size

Chara,
a common stonewort

SOME COMMON FUNGI

Black Rot on grapes

Green Mold on oranges

Bacterial Wilt on tobacco

Corn Smut on ear of corn

Shelf Fungi on a stump

Cup Fungi on rotting wood

Water Mold on living minnow

FUNGI, particularly mushrooms, are most conspicuous in their spore-producing or fruiting stage, often seen after the mycelium has spread widely through soil, wood, or a host.

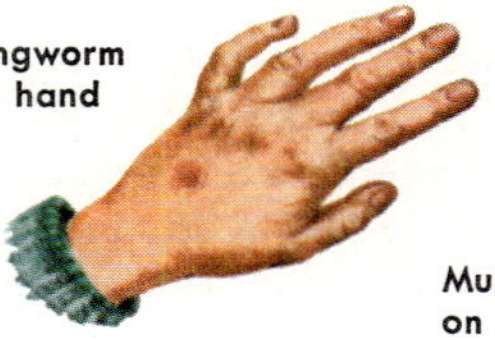

Ringworm on hand

Mushrooms on soil

FUNGI

The term fungi is commonly used for several groups of thallus plants that, lacking chlorophyll, subsist either as parasites or as saprophytes. It is more a general term than a botanical one. Bacteria, for example, are sometimes placed with blue-green algae rather than with fungi groups. But for convenience, in this book bacteria and slime molds are treated with the fungi. The chart on pp. 6-7 shows a possible evolutionary relationship among all these groups of non-flowering plants.

The majority of non-flowering plants are classified as fungi. In most, the vegetative body consists of fine threads, or hyphae, that, in a mass, form the mycelium. The mycelium of mushrooms grows underground; in parasitic fungi, it invades the host. Fungi produce asexual spores. Some swim by means of flagella; other non-motile kinds may form at the ends of hyphae or in spore cases. Spores are distributed by wind or water.

Most fungi also reproduce sexually. The process in some groups is complex. In these groups the contents of certain cells fuse, but not their nuclei. These new cells with two nuclei form fruiting bodies of various types. In the fruiting bodies the nuclei eventually fuse, then divide, and produce spores, each with only one nucleus. Cells of the mycelia that grow from these spores also contain only one nucleus.

Fungi and bacteria are responsible for many infectious diseases of animals and plants. Rot of wooden structures and of fresh fruit and vegetables; the mildewing of paint, clothing, and walls; and the spoilage of meats are all due to fungi or bacteria. But these simple plants also play an important role in the balance of nature. When plants and animals die, they would never decay were it not for fungi and bacteria that break down organic compounds and return them to air and soil.

Yeasts, bacteria, and other fungi cause fermentation by breaking down sugars and starch and producing alcohol. Yeasts also make dough rise. Blue molds spoil oranges but give blue cheeses their flavor, and related molds furnish penicillin and other antibiotics.

On the following pages are the five natural groups here called "fungi." These are bacteria (Schizomycetes), p. 27; slime molds (Myxomycetes), p. 28; algal fungi (Phycomycetes), p. 32; sac fungi (Ascomycetes), p. 34; and basidium fungi (Basidiomycetes), p. 40. The group called Imperfect Fungi (p. 92) includes those known only by their asexual stage.

EDIBLE MUSHROOMS have been prized since the days of the Greeks and Romans. Because of their exquisite flavor (they have little food value), people are anxious to try them. Two groups of fungi—Ascomycetes and Basidiomycetes—contain the common edible species. Many poisonous species, called toadstools, resemble or are closely related to edible species, even occurring in the same genera. Two species of *Amanita* are as hard to separate as two species of hawthorn—yet one *Amanita* is deadly and the other excellent eating. Nearly every group of the fleshy fungi includes one or more kinds either questionable or poisonous. Positive identification before eating is essential.

This book contains only a sample of the thousands of kinds of fleshy fungi. Many kinds, both edible and poisonous, are *not* included. Do not judge edibility by this book alone. Use more detailed references, and seek expert advice.

Hay bacilli

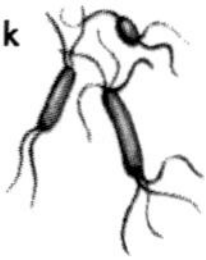

Blue-milk bacilli

Tetanus bacilli

BACTERIA

Diphtheria bacilli

Bacteria (Schizomycetes), the simplest plants, are all microscopic. They are sometimes classified with the blue-green algae, as both lack the well-defined nucleus found in the cells of other plants and animals. Bacteria reproduce by fission—one cell simply splitting into two. Under favorable conditions, bacteria increase at a prodigious rate. Most are harmless. *Sarcina,* for example, lives and multiplies on the human skin, remaining active even after repeated washing. Some are useful and some are pathogenic, causing such diseases as tuberculosis, diphtheria, cholera, tetanus, and anthrax. Bacteria also cause fire blight, crown gall, soft rot, and other plant diseases.

Best known of the bacteria are those that are rod-shaped (bacilli). These may form resting spores resistant to heat and drying. Other bacteria may be small spheres (cocci), occurring singly or in clumps or strings. Spiral types (spirilla) are less common. Some bacteria move by means of flagella.

Pneumonia cocci

Scarlet Fever cocci

Sarcina

Spirillum

Syphilis spirochaete

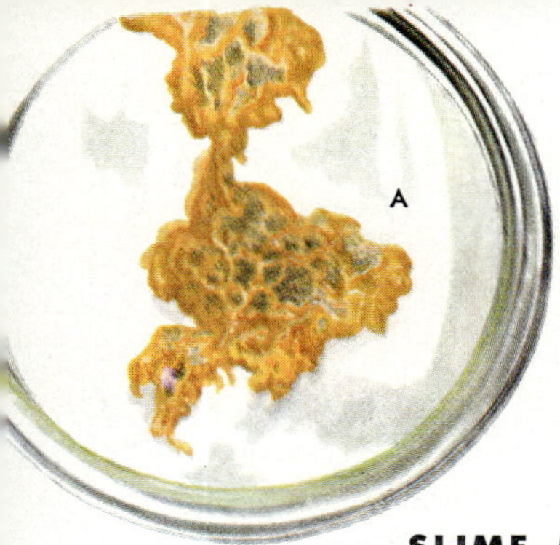

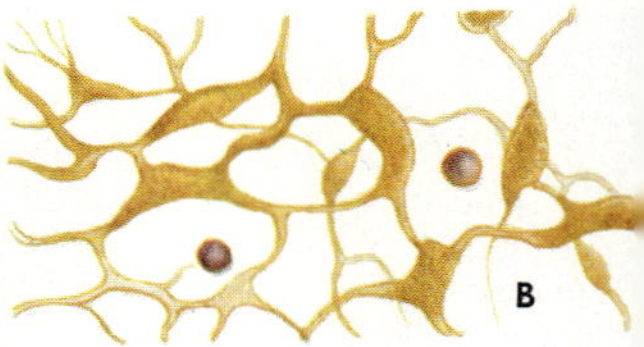

PHYSARUM plasmodium, in a petri dish (A), is magnified (B) to show veinlike branches of streaming protoplasm.

SLIME MOLDS

Slime molds (Myxomycetes) creep or flow along, feeding on bits of organic matter. Hence these strange organisms are sometimes considered to be amoebas rather than plants. But because slime molds also have a stationary, plantlike stage in which they produce spores, they can be grouped with the fungi.

A slime mold's jelly-like vegetative body, or plasmodium, is seldom seen, for it remains hidden in rich, loose soil, under decaying logs or leaves, or in other dark, moist places. This naked mass of protoplasm contains numerous nuclei but no cell walls. It moves like an amoeba, the protoplasm streaming internally first in one direction and then in another.

SERPENT SLIME MOLD has cordlike yellow sporangia that appear on decaying logs. Worldwide in distribution but most common in the tropics.

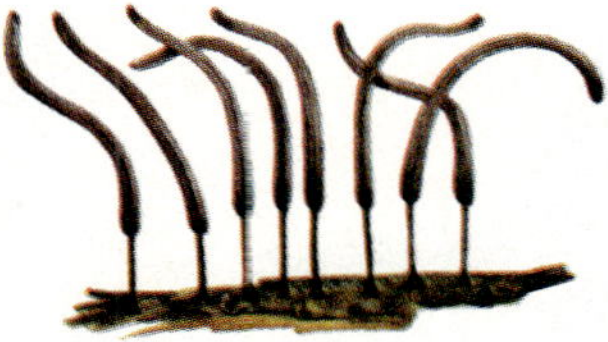

DEEP-BROWN STEMONITIS grows on decaying wood. The hairlike stalks extend up through the brown sporangium. They stand up stiffly, often in dense growths. Ht. 0.5-1 in.

WOOD-LOVING LYCOGALAS, common on dead wood, were once thought to be small puff-balls. The young, white to coral sporangia become tan or gray as they mature. Ht. 0.3-0.8 in.

A plasmodium that dries up will continue growing if moistened. A small piece broken off the original will grow into a new vegetative body, even after a long dormant period. A mature plasmodium contracts and forms, or converts, into fruiting bodies, or sporangia, each containing many spores. Sporangia appear on rotted wood, among decaying leaves, or even in grass. They occur in a variety of colors, but each species has a characteristic design. Some are on long stalks, made of waste materials; others are squat. Most are so small that their structure can be seen in detail only with a lens. Wind-carried spores from the sporangia germinate and eventually form a new plasmodium.

BASKET SLIME MOLD produces tiny, dark-brown sporangia like open-woven baskets atop slim stalks. The spores are inside the basket. Ht. 0.2 in.

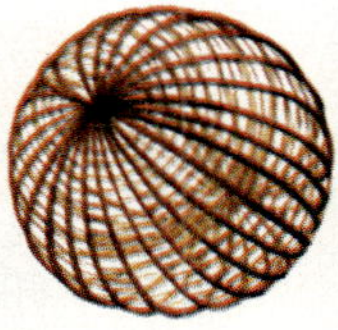

enlarged top of sporangium

SLIME MOLDS

FULIGO, largest slime mold, has stalkless, lime-encrusted, cushioned sporangia. They grow on the ground and are at first light tan, darkening to purple as lime breaks away and spores mature. Dia. 2-8 in.; ht. 1-2 in.

CRYSTALLINE DIDYMIUM is usually stalked. The iridescent surface of the sporangium is partially covered with minute lime crystals. This species has a number of varieties. Diameter of sporangium less than 0.04 in.

GRAY PHYSARUM has no stalk and is covered with lime. Spores are purple-brown. Occurs on manured ground, on rotting wood, in dead leaves. L. 0.4 in.

GREEN PHYSARUM varies from green through yellow to rusty orange. The covers of its stalked and nodding sporangia form irregular networks when they split. Ht. 0.2 in.

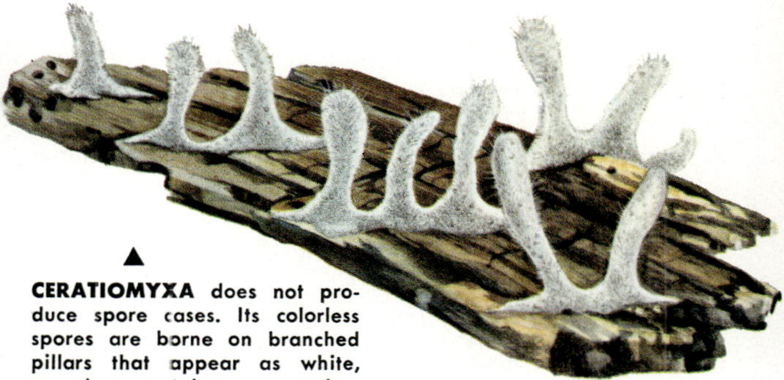

CERATIOMYXA does not produce spore cases. Its colorless spores are borne on branched pillars that appear as white, powdery patches on rotting wood. Common after summer showers. Ht. 0.08-0.4 in.

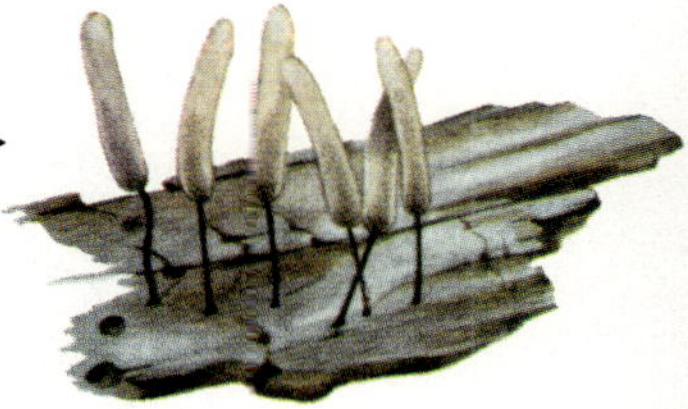

ASH-GRAY ARCYRIA has a short stalk, its outer covering soon disappearing except near base. The sporangium usually tapers upward and is sometimes globular. Network among spores is spiny or warty. Ht. 0.02 in.

VARIABLE TRICHIA, quite variable in form, is usually without a stalk (sessile). The sporangia, often crowded together, are yellow and shining. Ht. 0.4 in.

COMMON BADHAMIA is one of a number of species having lime-covered spore cases—some stalked, others sessile. In mass, spores purple-black; brown by transmitted light. Dia. to 0.04 in.

ALGAL FUNGI

Algal fungi (Phycomycetes)—about 1,500 species—are true fungi. All are microscopic. They lack crosswalls in their hyphae, and in this they differ from other fungi. As a result, the mycelium is made of branched hollow tubes containing many nuclei.

Algal fungi are the cause of blights and other plant diseases, with major economic and social consequences.

ASEXUAL reproduction involves formation of several kinds of spores. The wind-carried spores of Bread Mold develop in a stalked spore case, or sporangium (1). The zoospores of Water Mold are produced in a sporangium at the tip of a hypha (2). They have flagella and swim away from the parent before germinating. The spores of other algal fungi are formed by the breaking off or segmenting of a hyphal tip. Each portion becomes a sporangium, which may divide into zoospores (3) that later lose their flagella and develop into hyphae. Or a sporangium may germinate directly into a new hypha (4).

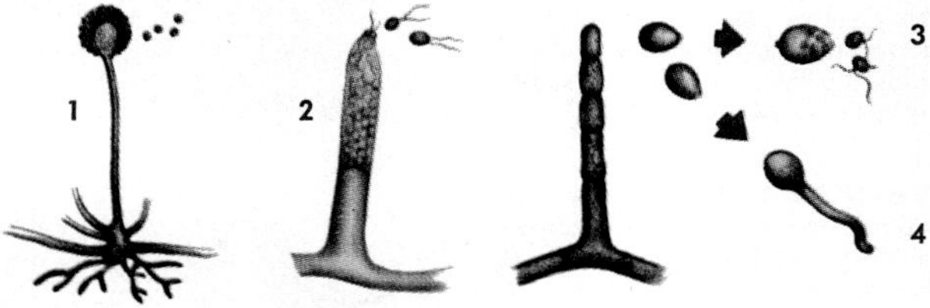

SEXUAL reproduction with similar cells (below, left) results in a resting-stage zygote. In other algal fungi (below, right) nuclei of cells from male organ (antheridium) fuse with eggs in female organ (oogonium). Zygotes germinate later.

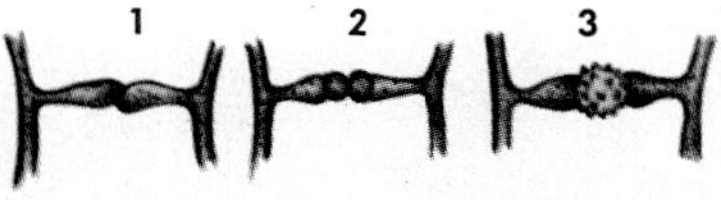

Branches extend (1), gametes fuse (2), and a zygote forms (3).

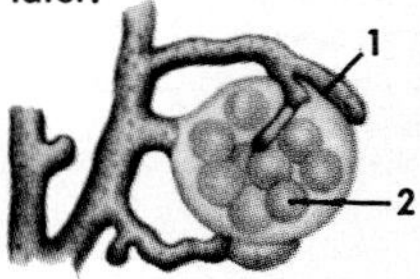

Nuclei in tube from antheridium (1) fertilize eggs in oogonium (2).

The great Irish famine of 1845-1846, caused by the Late Blight of potatoes, led to mass migration from Ireland. A downy mildew threatened the grape crop and wine industry in France until the development of chemical fungicides that brought it under control.

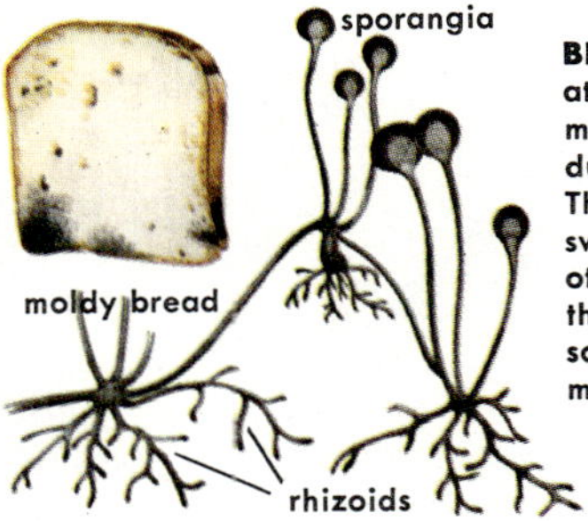

BREAD MOLD inhibitors added at bakeries delay growth of molds on bread and have reduced this once-serious problem. The same mold causes rot on sweet potatoes, strawberries, and other fruits and vegetables. But these common molds are also a source of cortisone and other medicines.

WATER MOLDS grow either in water or on damp soil. One kind appears as white, cottony masses on fish where scales have been damaged. Water Mold reproduces both asexually and sexually, shown here and on p. 32.

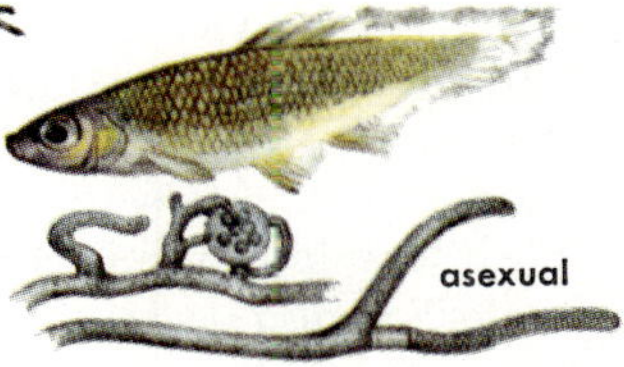

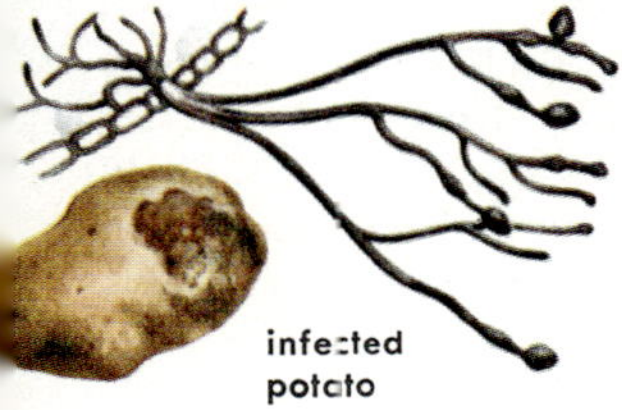

DOWNY MILDEWS cause Late Blight of potatoes and tomatoes and bud rot of coconuts. They also infect grapes, grains, tobacco, and other crops. Sporangia are seen as downy white, gray, or purplish patches.

WHITE RUST is a mildew that forms on leaves of cabbage-family plants. Blisters appear as the mildew reaches fruiting stage. When they break, white masses of sporangia are released.

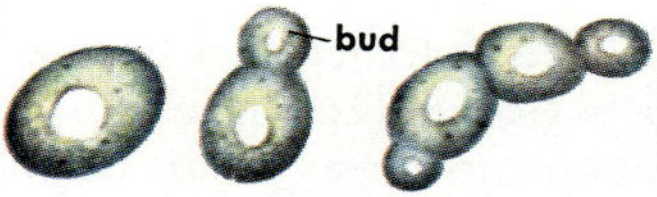

reproduction by budding

YEASTS are microscopic, one-celled plants that multiply rapidly as they ferment sugar or starch. Reproduction is mainly by budding but occasionally spores are formed in a thin sac.

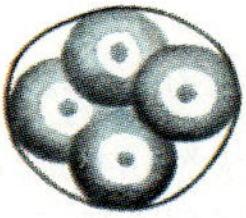

ascus with ascospores

commercial yeast

SAC FUNGI

Sac fungi (Ascomycetes) number about 40,000 species, most of them so small that they are rarely noticed. Sac fungi include the yeasts and the blue and green molds often seen on citrus fruits, jellies, and leather. Most reproduce asexually by means of spores (conidia), which form in chains at the ends of hyphae. Sexual reproduction also occurs, in the morels, truffles, and some other genera, by means of large, conspicuous fruiting bodies. Within these bodies small sacs (asci) are produced, each containing ascospores—usually eight. The hyphae have crosswalls that may separate cells with two nuclei or areas with even more nuclei. Sac fungi usually grow on dead wood or soil, though a number are parasites and cause plant diseases.

EDIBLE SAC FUNGI include the prized truffles and morels. Truffles are especially esteemed in Europe. In North America they are not common, hence rarely collected. Morels of the genus *Morchella* are edible and flavorsome, but species of the closely related *Verpa* are poisonous to some people. A False Morel *(Gyromitra* or *Helvella esculenta)* is credited with several lethal cases of poisoning. For eating, be certain that you collect only species positively identified as edible.

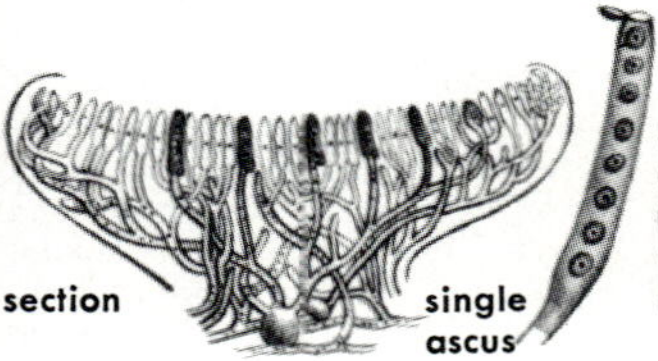

Section through fruiting body of a cup fungus (p. 37). Single ascus (enlarged) showing lid, which flops back at maturity, releasing eight ascospores.

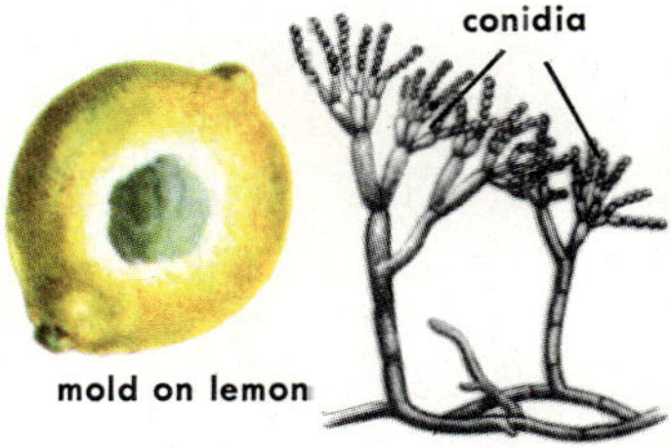

PENICILLIA are among the best known of the green molds. Species shown grows on citrus fruits and was the first source of the antibiotic penicillin. Various species of *Penicillium* may have blue, green, gray, or yellow spores. They are present almost everywhere and are also used to ripen Camembert, Roquefort, and other blue cheeses.

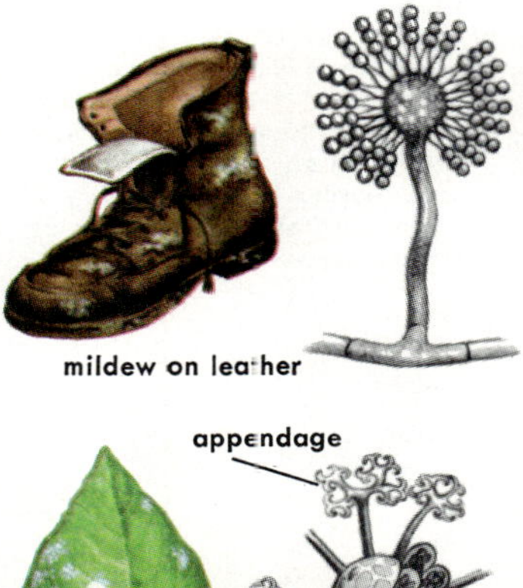

ASPERGILLI, molds closely related to *Penicillium,* may have black, green, blue, gray, yellow, or brown spores. Many organic acids are obtained industrially from these molds, which also cause mildew on walls, leather, and fruits. Respiratory diseases of birds and other animals may also be due to species of *Aspergillus.*

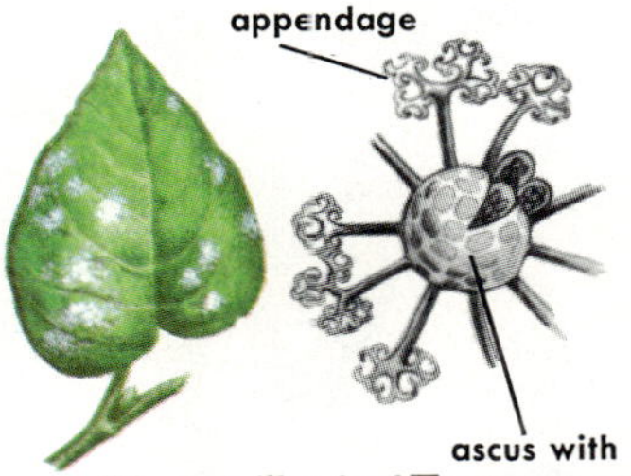

POWDERY MILDEWS are parasites of many plants. Asexual spores form a white powder that gives the name to the disease. Fruiting bodies are very small, and appear as tiny black dots on leaves. Characteristic outgrowths on the fruiting bodies are used in identification. Fruiting body contains one to several asci, three of which are exposed in illustration at right.

Common Morel
Dia. 1-2 in.;
ht. 2-5 in.

MORELS are a famous group of edible mushrooms fairly abundant in temperate regions. They are found in the spring in open woods and along stream banks. All are stalked, with pitted, spongy, rounded, or conical heads, varying in color from tan to brown or gray. All species of *Morchella* are edible. The Common Morel is shown here. *Verpa,* very similar but with cap free at base and loosely attached, may possibly cause illness.

FALSE MORELS have folded, chestnut to coffee-brown caps, often brainlike in appearance. The interior of the cap is white, uneven, and hollow. False morels are found on sandy soils during wet seasons. The Brain Gyromitra, illustrated, is known to be poisonous, sometimes lethal.

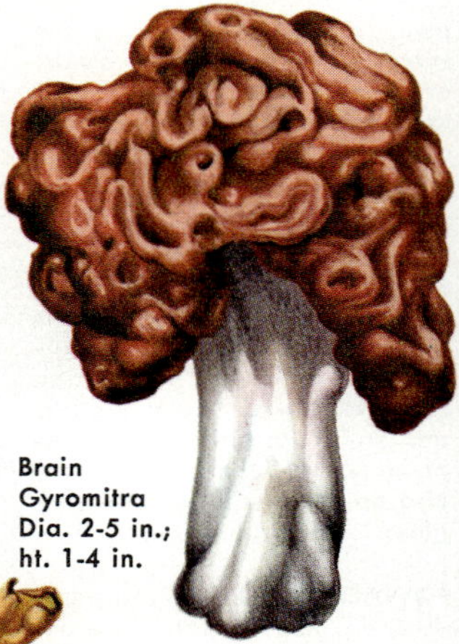

Brain
Gyromitra
Dia. 2-5 in.;
ht. 1-4 in.

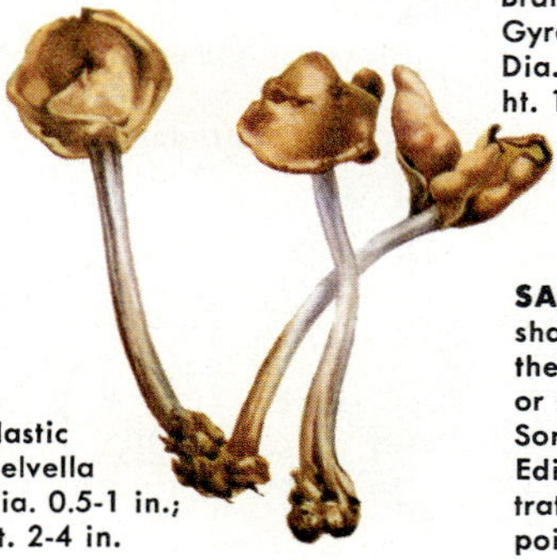

Elastic
Helvella
Dia. 0.5-1 in.;
ht. 2-4 in.

SADDLE FUNGI have saddle-shaped or irregular caps above their thin stalks, and vary in color from white to pink to brown. Some have an offensive odor. Edibility of Elastic Helvella, illustrated, is questionable; possibly poisonous.

TRUFFLES include several dozen species, mainly in two genera (*Terfezia* and *Tuber*). They are widely distributed in open woods, where they grow underground on roots of trees. In France and Italy, dogs and pigs are trained to dig them up. As a delicacy, no other mushrooms are more esteemed than the Summer Truffle and related species.

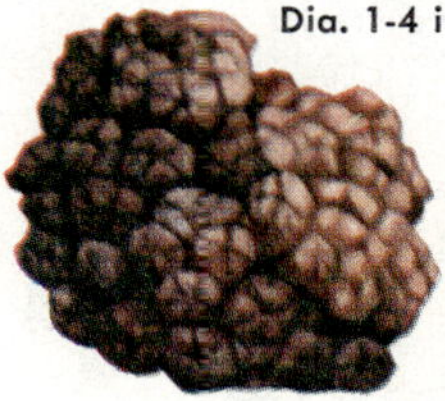

Summer Truffle
Dia. 1-4 in.

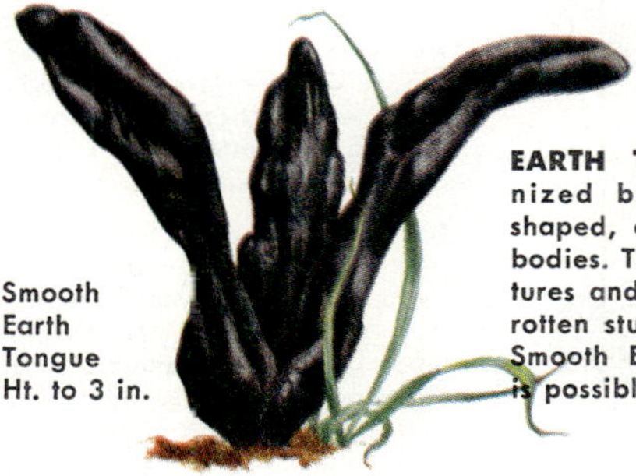

Smooth Earth Tongue
Ht. to 3 in.

EARTH TONGUES are recognized by their black, club-shaped, often flattened fruiting bodies. They grow in damp pastures and in the woods, on very rotten stumps, or in rich humus. Smooth Earth Tongue, pictured, is possibly poisonous.

CUP FUNGI, below and p. 38, are a large group of fleshy fungi, conspicuous though small. Usually not eaten. Orange Cup is a large, irregular, fragile species, found on the ground in damp woods. A smaller species, the Red Cup Fungus, has a bright-scarlet cup with a woolly, white or pink exterior. It grows chiefly on dead wood, sometimes half-buried in humus.

Orange Cup Fungus
Dia. 1-5 in.; ht. 0.5-1 in.

Red Cup Fungus
Dia. 1 in.; ht. 1 in.

Fire fungus growing on burned-over ground

CUP FUNGI

FIRE FUNGUS is common on burned-over ground or sites of old campfires, although occasionally found elsewhere. Large masses of this very small cup fungus are conspicuous because of the white cottony layer around the red or yellow disks. Dia. of cup 0.04-0.12 in.

cup enlarged

HELOTIUM is one of the many tiny cup fungi found on old stems and leaves. Most are yellow, white, brown, or purple. Some are stalked, others sessile. The many species are all similar in appearance, making them difficult to distinguish. Dia. of cup 0.04-0.12 in.

(1) cup stage on decayed peach

(2) brown rot stage on peach

MONILINIA is a cup fungus that causes brown rot of fruits. Long-stalked cups (1) arise in spring from remains of fruits that have fallen from trees. Spores ejected from sacs on the cup surface are wind-borne to fruits where rot (2) develops rapidly. Cup dia. 0.25-0.5 in.

CORYNE is one of the few gelatinous cup fungi. Its closely crowded, flesh-red to purple cups are turban-shaped. The asexual stage, convoluted and gelatinous, is more common than the sexual stage. The spores are spread by moisture and wind-blown raindrops. Found on old stumps. Cup dia. 0.08-0.5 in.

CARBON FUNGI AND ALLIES

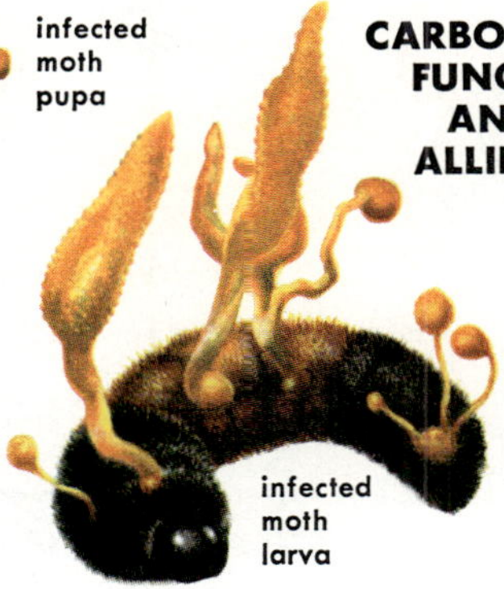

CORDYCEPS, the Vegetable Caterpillar, is parasitic on insects, attacking both larvae and pupae. Stalks with brightly colored, plumelike fruiting bodies grow from the victim. After a few weeks asci with asexual spores are produced on these growths. Ht. 1-2 in.

DALDINIA, a carbon fungus, is common on dead wood. The outer surface is purple at first, becoming black as spores mature. The inside is black, with concentric rings (see section at right). Small, flask-shaped, spore-producing bodies are embedded in this growth. Dia. 0.5-2.5 in.

DEAD MAN'S FINGER is a carbon fungus that produces asci-containing bodies on finger-like growths or clubs. Inside, the clubs are white. Common on dead wood (sometimes buried), it causes root rot of apple and other trees. Related species are red. Ht. 1-3 in.

ERGOT FUNGUS infects rye, wheat, and other grasses. It forms hard, black, sterile bodies in place of the cereal grains. When these fall to the ground, they produce stalked clubs that eject air-borne spores. Alkaloids from ergot are used to treat migraines and also in obstetrics. Ht. 0.3-0.5 in.

RUST, JELLY, AND EAR FUNGI

TRUE BASIDIOMYCETES

BASIDIUM FUNGI

Basidium fungi (Basidiomycetes), totaling some 25,000 species, form a group that includes many edible mushrooms (below). Reproduction in basidium fungi is complex, especially among the rusts and smuts. Several types of asexual spores are produced, and at some phase in the life cycle, all basidiomycetes produce basidiospores following sexual reproduction. Basidiospores are borne on club-shaped structures (basidia), each of which usually produces four spores. The color of these spores is an important identifying feature for some species of mushrooms and is best noted when the spores are seen in mass, as in a spore print (p. 10).

EDIBLE BASIDIOMYCETES share company with lethal species. The *Amanita phalloides* group (p. 52) accounts for about 90 percent of all the deaths due to mushroom poisoning. *Boletus* and *Russula* are also genera containing both edible and poisonous species. Puffballs in the genus *Scleroderma* are reportedly poisonous, though most puffballs are excellent eating.

If you are interested in mushrooms as food (and many are delicious), first read widely and study illustrations in a number of references. Collect with an expert if you can. Keep each species in a separate bag or container, and discard those that are wormy or spoiled. Do not eat even a small amount if you have the slightest doubt about identification. As the U.S. Public Health Service says, "Certainly, wild mushrooms should not be used for food unless their identity and lack of toxicity have been established beyond a shadow of a doubt."

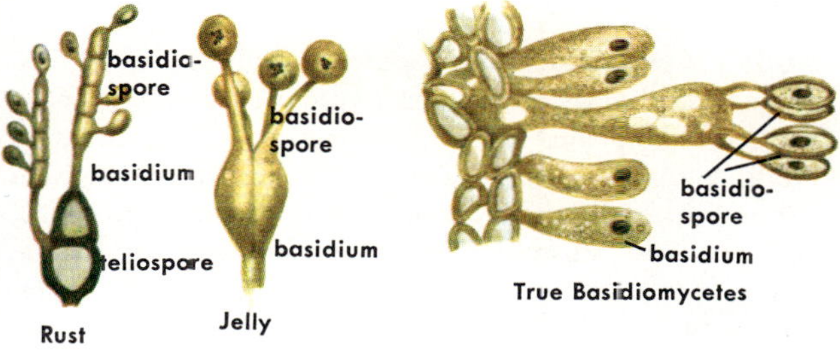

TYPES OF BASIDIA

Basidiomycetes are divided into two large groups, as shown on top of p. 40. In fungi of the first group, which includes the rusts and smuts, and the jelly and ear fungi, the basidium is divided into cells. The complex life histories of rusts and smuts were only determined after years of painstaking research.

In the second group, the basidium is not divided. These are the true basidiomycetes, with fleshy, papery, or woody fruiting bodies. True basidiomycetes are further divided into two groups—the Hymenomycetes, p. 46, and the Gasteromycetes, p. 86. Examples of both groups are illustrated below.

Examples of Hymenomycetes (H) and Gasteromycetes (G)

RUSTS AND SMUTS are parasites of higher plants. They spread widely and can impair the economy of a region. Wheat Rust can reduce the crop 50 percent or more. Blister Rust has killed large stands of valuable White Pine. Scores of plants, from asparagus to roses, are attacked. Rusts cause serious diseases of coffee.

A rust is identified by the host it infects, its life cycle, and especially the form of its resting spores. Rusts may be very specific in the plants they attack. Some infect closely related species; others a single species, and some only a variety of a species; other varieties are immune. Because of the damage they cause, smuts and rusts have been fully studied.

CORN SMUT attacks corn wherever it grows. All parts of the plant are susceptible. Spores land on corn and germinate. The hyphae penetrate the host, forming "boils" that consist of dark mycelium covered by a white sheath of corn tissue. Corn Smut basidiospores are of two strains, or sexes, as are the hyphae that grow from them. When these fuse, the smut forms numerous thick-walled spores that fill the boil as a dark "smutty" mass. When the boil bursts and liberates them, these spores germinate and produce basidiospores that may bud again and again, increasing the total number of spores many times. Carried by wind, basidiospores land on corn plants, repeating this cycle every few weeks and spreading the infection. In winter the resting spores remain dormant in soil or litter until spring.

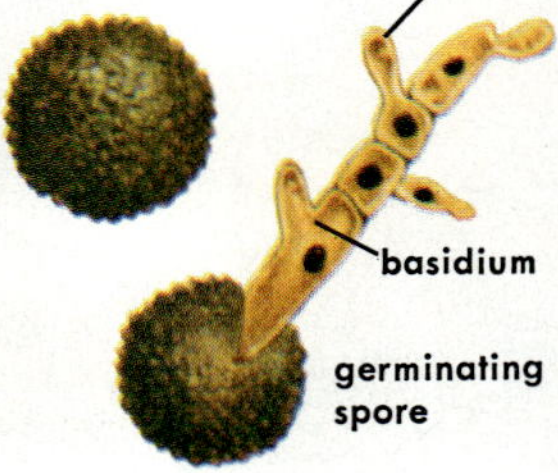

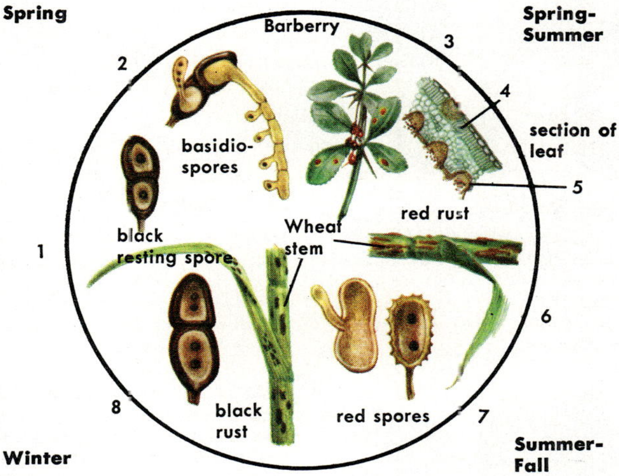

WHEAT RUST has a many-phased life cycle. Thick-walled resting spores (1) winter in soil or stubble. In early spring they produce basidiospores (2) of two different strains, or sexes. If these wind-blown spores reach barberry leaves, they germinate (3). Later, yellowish-red bumps appear on top of leaf (4). These release spores and a sticky nectar that attracts insects, which pick up spores and may also bring spores of an opposite strain. Hyphae and spores of opposite strains must fuse before spores are produced in clusters on undersurface of leaf (5). Some spores are blown to wheat plants, where they germinate on stems and leaves (6). Soon reddish streaks or patches of spores appear. These spores (7) may infect other wheat plants, repeating cycle. In fall, black resting spores are again produced (8).

Rusts have complicated life cycles, some with as many as five spore stages. Many rusts require two different hosts; some need only one (p. 44).

Smuts have simpler life histories. They cause blisters and swellings on plants they infect or form black scabs or yellow spots on the leaves. Sometimes spores of smuts stick to seeds and germinate with them.

CEDAR APPLE RUST requires two hosts. Spores formed in structures on the underside of apple leaves infect cedars and junipers. There they form fleshy galls, or "apples," that produce two-celled resting spores. These spores are borne back to apple leaves by wind. Sexual reproduction involves spores formed in structures on top of the apple leaves. Removing cedars controls the disease, as does planting of rust-resistant apples.

BEAN RUST completes its life cycle on one host, attacking only bean plants. In summer reddish-brown spore patches appear on underside of bean leaflets. Dark, one-celled resting spores are formed in fall and winter.

WHITE PINE BLISTER RUST forms yellow, followed by white, blisters on pine twigs. Spores from white blisters infect gooseberries or currants. There the fungus forms yellowish spores that reinfect nearby bushes. Other spores again infect pine.

ASPARAGUS RUST is a short-cycle rust, producing three types of spores but completing its life cycle on asparagus. Development of rust-resistant varieties has limited damage to crops. Pustules of resting spores on stems are illustrated.

JELLY FUNGI are usually gelatinous, especially when wet. When dry, the colorful fruiting bodies are rigid and horny but may regain their original form and jelly-like texture when moistened. Some common jelly fungi have pear-shaped basidia divided longitudinally, with the basidiospores produced at the top of each narrowed neck (p. 41). In ear fungi, four-celled basidia cover the concave surface of the irregular fruiting body.

Clammy Calocera

CALOCERAS resemble coral fungi (p. 46) but are tough and gelatinous. When fresh, their surface is slimy; horny and more orange-colored when dry. Common on stumps and dead twigs of conifers. Ht. to 2.5 in.

Ear Fungus

EAR FUNGI superficially resemble human ears in shape and foldings. They grow to 3 or 4 in. in diameter, shrinking and becoming hard when dry. When wet, texture is rubbery. Edible.

TREMELLAS, a group of about 40 species, are very jelly-like. They grow on trees or dead wood. The folded and twisted fruiting bodies of different species may be white, orange, mulberry, or yellow. No current reports on edibility. To 3 in. wide.

Orange Tremella

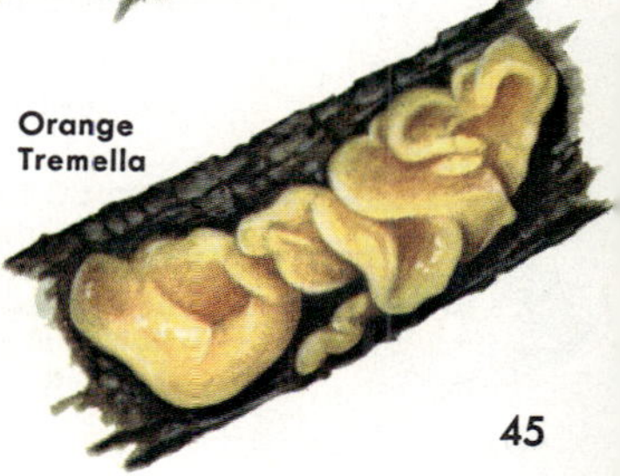

TRUE BASIDIOMYCETES are the best-known basidium fungi. The group includes most of the mushrooms. All have a single-celled basidium that produces four basidiospores. The fruiting bodies on which basidia form are often large and conspicuous. Two more or less natural groups are the Hymenomycetes and the Gasteromycetes (p. 86). For typical mushrooms, the width of cap is given followed by height of stem.

HYMENOMYCETES

Hymenomycetes are a large, important, and heterogeneous group. Their basic characteristic is that each exposes its hymenium (the fertile layer of basidia) before the spores are mature. The Hymenomycetes include: the coral, tooth, and leather fungi (below and p. 47); the pore fungi (pp. 48-50); and the gill fungi (pp. 51-85).

CORAL, TOOTH, AND LEATHER FUNGI have basidia exposed on toothlike structures (on branches or below caps) or on the smooth, leathery or fleshy surfaces of fruiting bodies. Some species are edible, others are woody or tough, and a few are known to be poisonous.

STRAIGHT CORAL FUNGUS is an upright fungus with stiff ascending branches, often compact; yellow to buff, occasionally violet-tinged. Grows on wood; tough and bitter. Ht. 2-4 in.

ASHY CORAL FUNGUS is white to gray, darkening with age. Erect, branching, with stalks fused at base, it is fleshy, fragile, and variable in form. Excellent eating. Ht. 1.5-3 in.

HEDGEHOG HYDNUM hangs from rotting tree limbs. The many-branched fruiting body is at first white, turning yellow or tan with age. It is a mass of long, partially fused, rather tough "spines." This striking fungus is edible, as are most *Hydnums*. Dia. 2-4 in.; ht. 2-4 in.

REPAND HYDNUM usually has an off-center stem. Smooth cap, like a gill mushroom's, has numerous white teeth below. Cap is thick; stem short. Edible. Dia. 1-4 in. (or more); ht. 1.5 in.

HORN OF PLENTY is trumpet-shaped, with a wavy margin on the cap. Dull brown and scaly when dry, becoming almost black when moist. Often found in great numbers on the ground, in woods. Difficult to see, as it blends with dead leaves. Edible. Dia. 2-4 in.; ht. 1-3 in.

HAIRY STEREUM is common all year, growing either as leathery, inedible brackets or as funnels on stumps, logs, or decaying wood. Note the zoning on the hairy upper surface. The smooth, bright-yellow, spore-bearing layer below turns gray with age. About 4 in. wide.

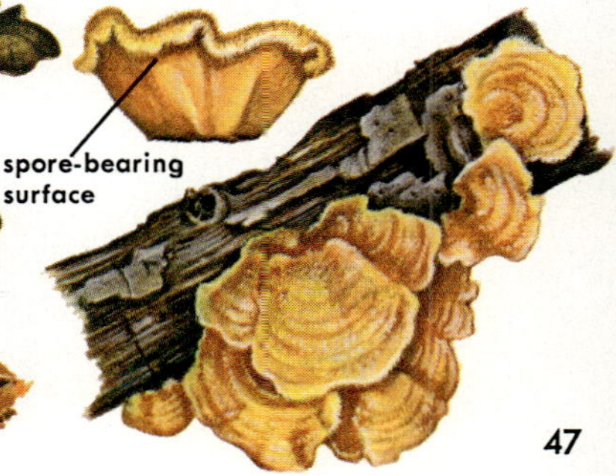

EDIBLE BOLETE's cap resembles a toasted bun. Its cream-colored pores or tubes eventually become greenish. Stem pestle-shaped. Grows in woods on ground. Edible. Dia. 4-6 in.; ht. 2-6 in.

LURID BOLETE has a bright-red to maroon pore surface. Tubes and inner, yellow flesh become blue when cut. A raised network of red lines covers stem. Poisonous. Dia. 4-8 in.; ht. 2-6 in.

PORE FUNGI are true basidiomycetes, with tubes or pores on the underside of the conspicuous fruiting body. Basidia in the pores produce spores when mature. Many common year-round species are found in this large group, which contains the polypores and boletes. Most of the polypores are woody, leathery, or papery, but the boletes are fleshy and soon decay.

Boletes, which grow on the ground, are stemmed fungi with pores instead of gills. The pores separate easily from the cap. Species are not easy to distinguish.

YELLOW CRACKED BOLETE has a soft, smoky-brown cap with yellow cracks around margin. Sulphur-yellow tubes turn blue. Grows in grass near woods. Edibility poor. Dia. 1-4 in.; ht. 2-4 in.

PINE CONE FUNGUS has surface tufts on the cap that resemble pine-cone scales. The pale-tan flesh turns red-brown, black when injured. Edibility poor. Dia. 2-4 in.; ht. 2-5 in.

SULPHUR POLYPORUS usually grows in overlapping shelves on living oaks or in clumps on fallen logs. Only the tips of young plants are eaten; other parts are too tough. 3-6 in. wide.

MULTI-ZONED POLYSTICTUS has a thin, leathery, beautifully zoned fruiting body. Concentric bands of varied dull colors mark the caps, which cluster on dead limbs. 1-4 in. wide.

If the flesh turns blue when cut or bruised, the mushroom may be poisonous. Several are bitter, but a large number are edible and prized by gourmets.

In polypores the pore layer is difficult to separate from the cap. These fungi usually form overlapping, fan-shaped brackets, or shelves, or clusters on tree trunks or dead wood. The mycelium may invade the heartwood of living trees, turning it to a brown, charcoal-like mass. New fruiting bodies may appear year after year in the same place.

RED POLYPORE is blood red but gradually fades to yellow and finally to white. This southern fungus, similar to the northern Cinnabar Polypore, grows on dead logs. Woody. 1-3 in. wide.

FIR POLYPORE, common on conifers, produces shell-shaped, flexible shelves. Gray above, lilac-colored below. Each pore opens from a spikelike tooth. Shelves 0.5-1 in. wide.

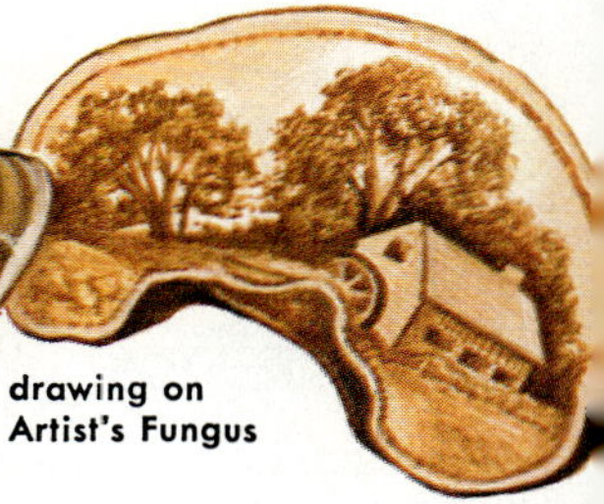

drawing on
Artist's Fungus

ARTIST'S FUNGUS grows as a bracket. The top is smooth, hard, and zoned. The tubes below are pure white but turn dark when injured. Some artists draw pictures on the flat tube surface. Found on oak and beech trunks. 8-24 in. wide.

section
through
fungus

RUSTY-HOOF FOMES is a woody, perennial shelf fungus that may grow for 35 years or more. Each year or so a new zone of tubes is added. This species was once widely used as tinder and punk. Found on beech and birch and a few other hardwood trees. 3-12 in. wide.

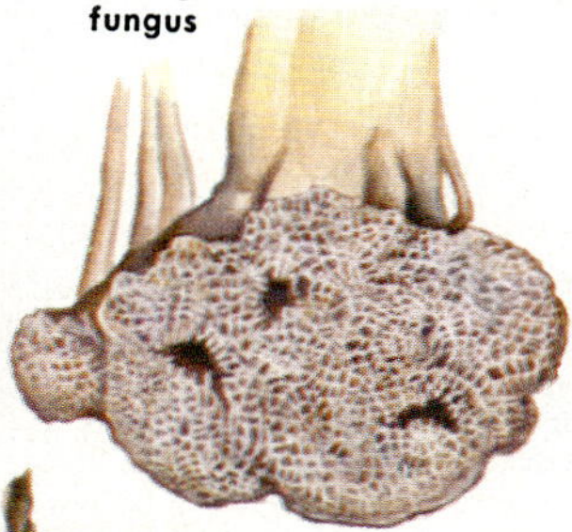

DRY-ROT PORIA grows flat against wood or hangs from timbers. This fungus does serious damage to construction wood, often destroying sills and beams, and the ribs and planking of wooden ships. 2-20 in. wide.

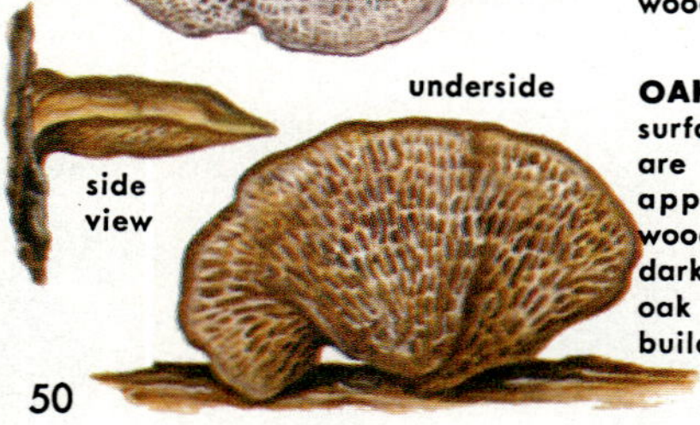

underside

side
view

OAK DAEDALEA has an undersurface like a labyrinth. Pores are irregular and long, at times appear almost gill-like. The woody, gray to brown brackets darken with age. Common on oak stumps; may infect wooden buildings. Shelves 3-8 in. wide.

GILL FUNGI typically have a distinct stem and cap, as shown below. On the underside of the cap are the plate-like gills that bear the basidia and spores. The gills, which radiate from the stem to the margin, may be broad or narrow, spaced close together or far apart. They may be forked, notched, or saw-toothed. The way the gills are attached to the stem, illustrated below right, is used in identification.

The mycelium of gill fungi grows in the ground or in rotted wood. Fruiting bodies appear as small buttons that enlarge and rupture. In the button stage, gill fungi could be mistaken for puffballs (p. 86). Many gill fungi are edible. Many are not, and a few kinds are deadly poisonous. Gill fungi are treated by genera, grouped according to spore color. Those with white spores are on pp. 52-76; pink, pp. 77-78; brown, pp. 79-80; purple, 81-83; black, 84-85. See page 10 for help in making a spore print.

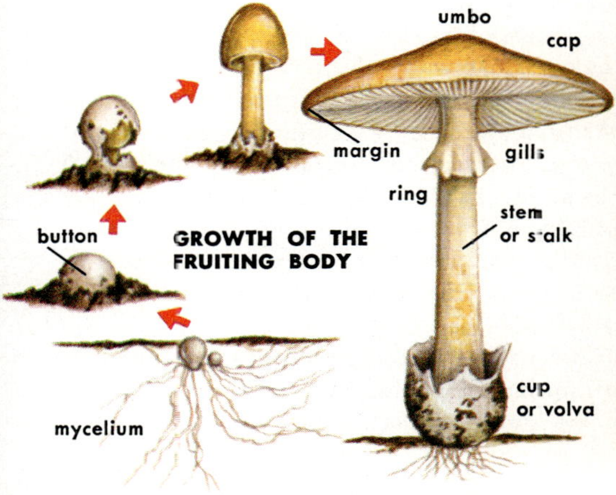

ATTACHMENT OF GILLS TO STEM

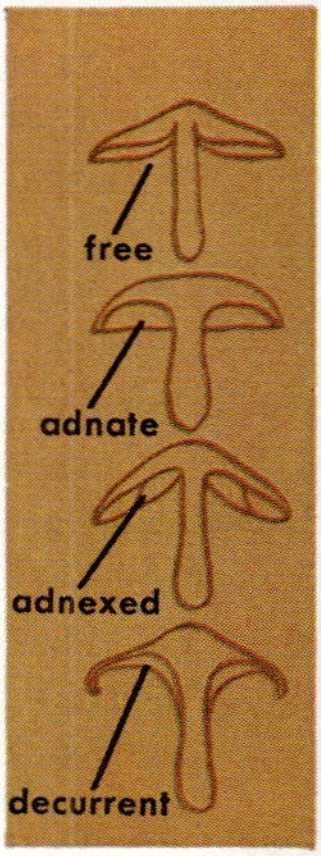

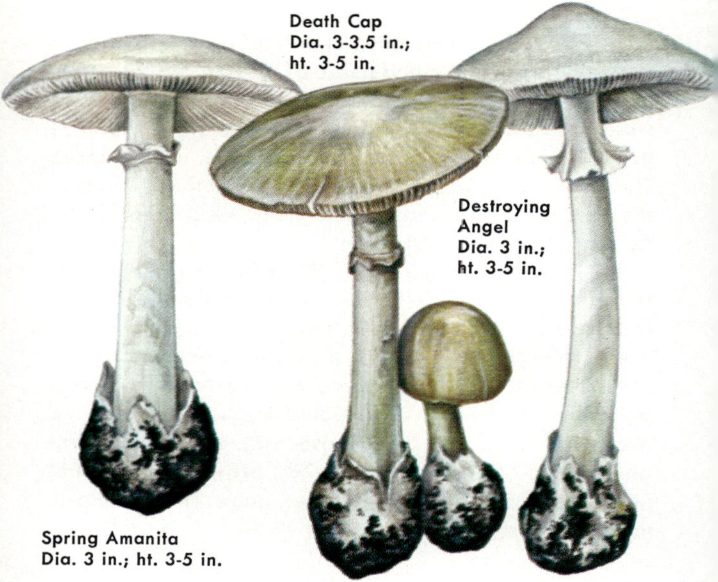

AMANITAS include the most poisonous mushrooms as well as edible species (p. 54). It is wisest to avoid all Amanitas, as a mistake within this group could be fatal. Those shown directly above account for over 90 percent of the deaths from mushroom poisoning. They are especially dangerous because no effects are noticed for 10 hours or more. By then it may be too late. The Fly Amanita contains a different poison, which takes effect quickly. Vomiting and drugs can help, so Fly Amanita need not be fatal. Small amounts of it are eaten ceremonially by primitive people.

Amanitas have white spores, a large cup (volva) at the base, a ring just below the cap, and free gills (not attached to the stem). Amanitas grow in open woods.

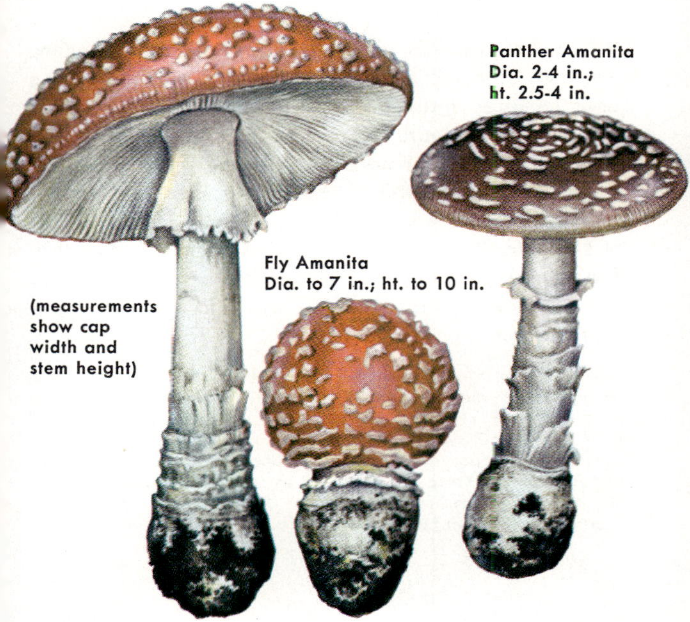

SPRING AMANITA, DEATH CAP, AND DESTROYING ANGEL are so similar that little is gained by trying to separate them. All are deadly poisonous. Color is not a guide. It varies from white (most common) to greenish, buff, or tan. All are found in open woods or clearings in summer and fall. The Spring Amanita has a slender stem; its cap may have a brownish center. The Death Cap has a smooth stem; the cap may be greenish. The Destroying Angel may have scaly stem; viscid cap has central swelling (umbo).

FLY AMANITA is also variable; color ranges from yellow to orange and red. Typical are white warts on the cap, a heavy stem, and a volva that is not pouch-like, as in species on p. 52. Found in open woods, often under conifers. Easily noticed because of its bright colors, it has been confused with the edible Amanitas (p. 54).

PANTHER AMANITA is dull yellow, gray, or brownish. The edge of the cap is furrowed; the volva reduced. Reported more poisonous than the Fly Amanita.

EDIBLE AMANITAS are not unique, for other genera also have both edible and poisonous species. Why one genus includes lethal and edible species is not understood. In the *Amanita* group at least five kinds of poisons are involved. Some are destroyed by heat; others are not. Those Amanitas that are edible (below) resemble poisonous species. The cap of the Blusher is like the Fly Amanita, and the volva of Caesar's is like that of the Destroying Angel. Experts suggest that these species be avoided to eliminate any chance of wrong identification.

BLUSHER resembles the Fly Amanita but is dull red-brown, becoming more reddish when cut or bruised. Grows in fields as well as in open woods in summer and fall. Volva not distinct; ring large and well developed. Dia. 3-7 in.; ht. 3-9 in.

CAESAR'S AMANITA has a smooth, shiny cap, strongly lined at the edge. Gills yellow and free; volva large, white, and cup-shaped. A fall mushroom of open woods, it has been prized for its fine flavor since Roman times. Dia. 3-7 in.; ht. 4-7 in.

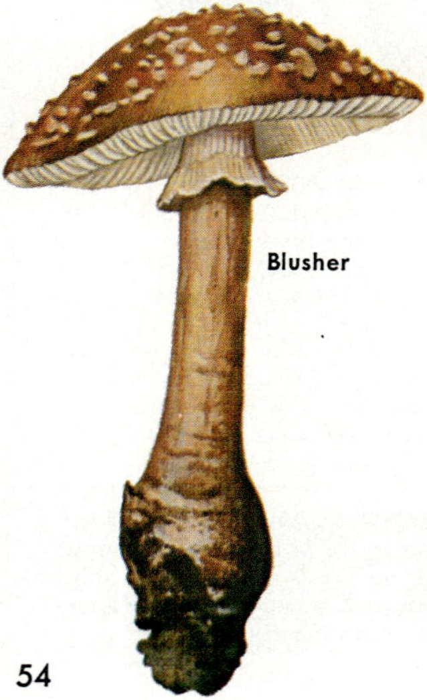

Blusher

Caesar's Amanita

AMANITOPSIS is a group resembling *Amanita,* with which it is classified by some authorities. Most species have a conspicuous volva, but they lack the ring that encircles the stem of the Amanitas. The gills are free, and the white spores are larger and rounder than those of Amanitas.

SHEATHED AMANITOPSIS varies from tawny to dark brown, mealy, or white. The volva forms a sheath around the base of the stem. The cap, moist to sticky in young specimens, is deeply striated along the margin. Sometimes an umbo develops, though the caps are usually flat. Grows in a great variety of habitats in cooler parts of North America and Europe. Edible. Dia. 2-4 in.; ht. 5-6 in.

VOLVATE AMANITOPSIS has a white cap that is hairy or scaly. Volva is large. In tests, guinea pigs and rabbits have been poisoned by this mushroom; no records of human deaths. Dia. 1-3 in.; ht. 2-3 in.

STRANGULATED AMANITOPSIS has a grayish-brown, warty cap. Volva usually torn into several "collars" that encircle lower stem. Edible. Dia. 3 in.; ht. 5 in.

ARMILLARIAS form a small genus of mushrooms that have a conspicuous ring, though this is occasionally small or even lacking. Spores are white, as in Amanitas, but Armillarias have no volva, and gills adhere to stem. Caps are smooth to slightly scaly.

HONEY ARMILLARIA is abundant in temperate areas, causing root rot of oaks and other hardwoods. Black strands of a luminous mycelium grow under the bark and spread underground to other trees. Honey Armillaria grows in dense clusters. Caps are honey yellow to reddish brown and usually scaly. Edible. Dia. 2-6 in.; ht. 3-6 in.

SLIMY ARMILLARIA, with white or grayish caps, grows on living or dead beech trees. Edible, but not desirable. Cleaning difficult, as dirt sticks to slimy caps. Dia. 1-6 in.; ht. 1.5-3 in.

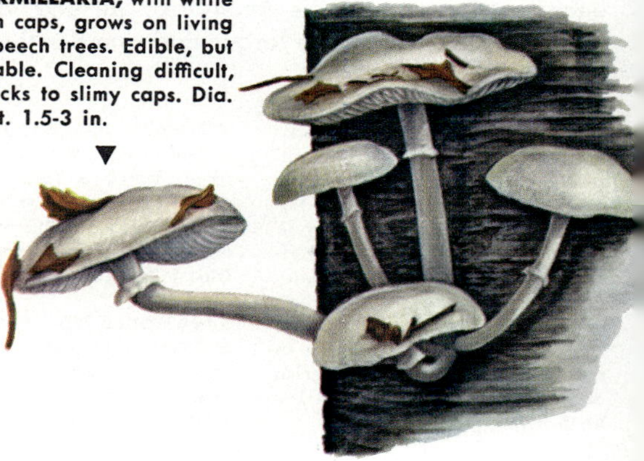

LEPIOTAS include some of the largest and most beautiful mushrooms as well as many small species. Eight species are illustrated on pp. 58-59. Lepiotas differ from Amanitas chiefly in the absence of a volva, although the base of the stem may be enlarged. The free gills produce white (occasionally pink or green) spores. Some species, including the Parasol Lepiota, often have a well-developed,movable stem ring. Smaller species may have an incomplete ring or none. Many develop an umbo. Most of the tested species of *Lepiota* are edible, but a few, such as Morgan's Lepiota, are known to be poisonous. Ordinarily Morgan's Lepiota is not fatal, but it can be dangerous, especially to small children. It upsets the stomach and lower digestive tract. Morgan's Lepiota forms a green spore print; edible Lepiotas give white spore prints.

Green spore print on white paper of poisonous Morgan's Lepiota (p. 58).

White spore print of Parasol Lepiota (p. 58) on dark paper.

MORGAN'S LEPIOTA is so much like the Parasol Lepiota that it is easily mistaken for it. Spore prints dull green (p. 57). Poisonous. Dia. 3-8 in.; ht. 3-8 in.

PARASOL LEPIOTA is edible but resembles poisonous Morgan's Lepiota. Stem has brownish scales and a movable white ring. Dia. 5-8 in.; ht. 6-11 in.

SHAGGY PARASOL also resembles Morgan's Lepiota but has larger, more shaggy scales and a white, smooth stem. Flesh of cap white; red when cut. Edible. Dia. 3-7 in.; ht. 3-9 in.

CRESTED LEPIOTA has rusty-brown scales toward center of its small, silky, white cap. Stem ring soon disappears. Has a radish odor when crushed. Possibly poisonous. Dia. 1-2 in.; ht. 1-2 in.

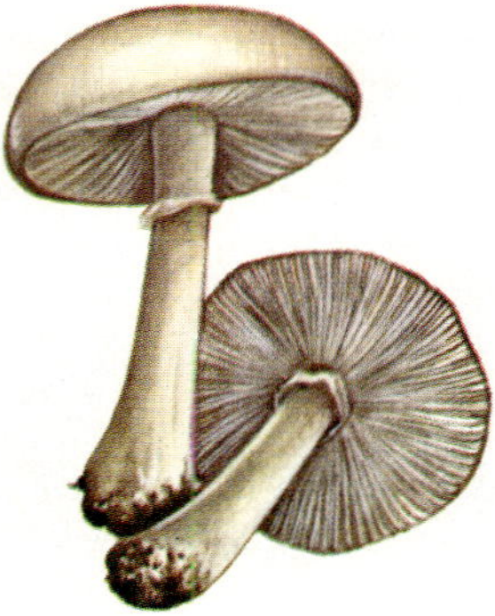

SMOOTH LEPIOTA has no scales. Cap and gills white. Gills age to smoky brown. Possibly poisonous; resembles Spring Amanita. Dia. 2-4 in.; ht. 2-3 in.

SHIELD LEPIOTA starts out bell-shaped and becomes flattened. Cap covered by brownish, felt-like layers that break up into scales. Found in woods. Edible. Dia. 1-3 in.; ht. 2-3 in.

AMERICAN LEPIOTA has reddish-brown scales on knobbed white cap. Stem thickens above base; ring often disappears. Edible. Dia. 1-4 in.; ht. 3-5 in.

YELLOW LEPIOTA often appears in greenhouses or among potted plants. Sulphur-yellow cap and stem are covered with fluffy scales. May be poisonous. Dia. 1-1.5 in.; ht. 1.5-3 in.

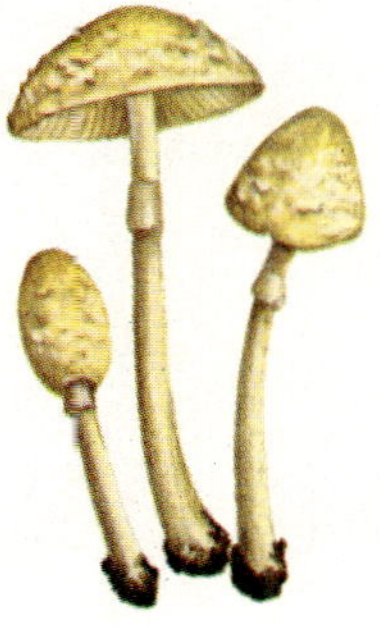

TRICHOLOMAS all lack ring and volva. Spores are white except in a few species where, in mass, they appear flesh-colored. These are sometimes put in another genus. Gills are adnexed (notched or curved upward just before attaching to stem, p. 51). Many are edible, but Leopard Tricholoma may be lethal.

LEOPARD TRICHOLOMA has a cap with white flesh and dark-gray scales. Grows under conifers in cool temperate areas. Poisonous. Dia. 3-6 in.; ht. 2-5 in.

SOAPY TRICHOLOMA has a soapy smell. Cap never sticky, becomes scaly in dry weather. Gills and flesh turn red when broken. Dia. 2-4 in.; ht. 2-4 in.

RED-HAIRED TRICHOLOMA has a dry cap covered with dark-red or purple, downy scales. This mushroom is common near pines. Dia. 2-6 in.; ht. 2-4 in.

Leopard Tricholoma

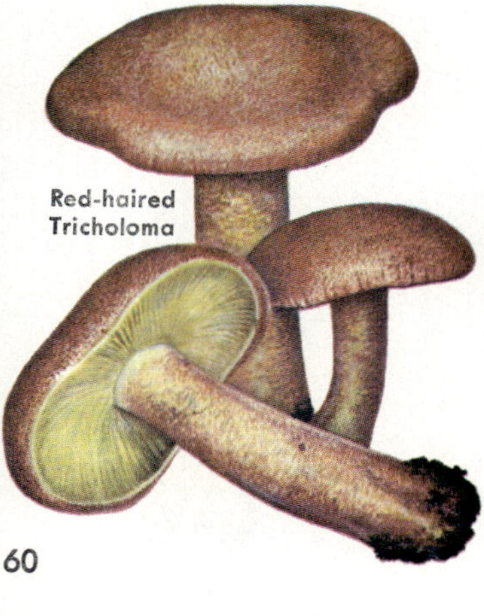

Red-haired Tricholoma

Soapy Tricholoma

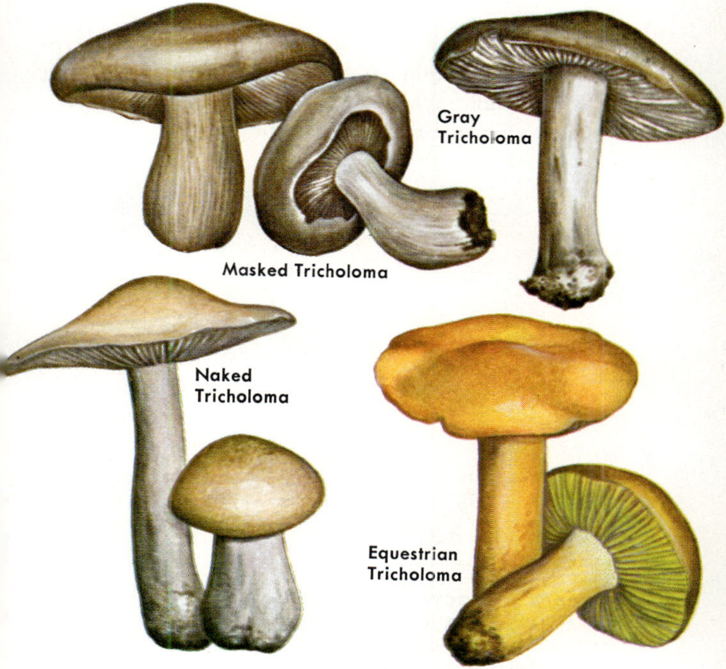

MASKED TRICHOLOMA, or Blewit, grows in pastures, often forming large "fairy rings." Sometimes purplish, especially when young, but usually becomes paie to buff. Broad gills gray or pinkish. Edible and very desirable. Dia. 2.5-5 in.; ht. 3 in.

NAKED TRICHOLOMA, or Wood Blewit, has a thin cap with a smooth, wavy margin. The flesh is lilac when young; gray or brown when older. Edible and common. Dia. 3-4 in.; ht. 2-4 in.

GRAY TRICHOLOMA is a variable, scaly-capped species that grows in clusters in pine and deciduous woods. This edible mushroom may appear in great numbers in late autumn, when most others have gone. Dia. 1.5-3 in.; ht. 2-3 in.

EQUESTRIAN TRICHOLOMA has reddish scales on cap; viscid when wet. Gills, sulphur yellow, do not change when cut. Grows in pine woods. Edible. Dia. 1.5-4 in.; ht. 1-3 in.

CLITOCYBES, similar to Tricholomas (pp. 60-61), have white spores, but gills are usually decurrent (extend down stem). Stem lacks a ring. Typically, cap is funnel-shaped. Species are difficult to distinguish.

YELLOW CLITOCYBE resembles Chanterelle (p. 76) but has orange-red gills that are crowded, thin, and sometimes forked. The slightly depressed cap varies from pale yellow to dark brown and almost black. May be poisonous. Dia. 1-3 in.; ht. 1-2 in.

ADIRONDACK CLITOCYBE, of northern forests, has a central depression in the wavy-margined cap. Edible, fine-flavored. Dia. 1-2 in.; ht. 1.5-3 in.

SWEET-SCENTED CLITOCYBE is edible, with anise odor and pleasant but strong flavor. Gills are white; greenish white when older. Dia. 2-3 in.; ht. 1.5-2 in.

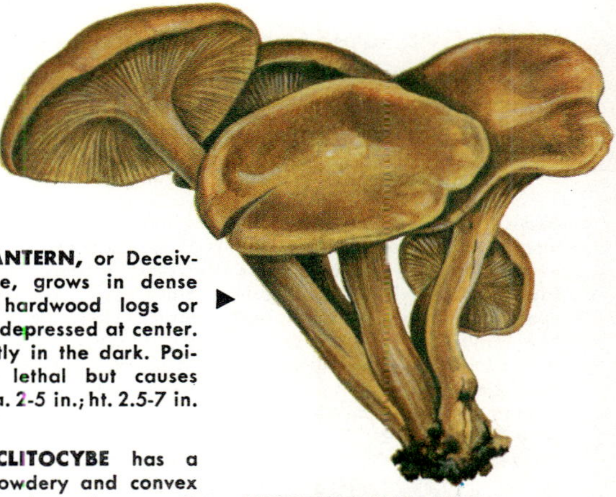

JACK-O'-LANTERN, or Deceiving Clitocybe, grows in dense clusters on hardwood logs or stumps. Cap depressed at center. Glows brightly in the dark. Poisonous; not lethal but causes vomiting. Dia. 2-5 in.; ht. 2.5-7 in. ▶

CLOUDED CLITOCYBE has a wide cap, powdery and convex at first but later becoming smooth and depressed. Stem thick. May form large clusters in deciduous woods. Can cause stomach upset. Odor rather disagreeable. Dia. 3-7 in.; ht. 3-5 in.

SUDORIFIC CLITOCYBE is gray to creamy white; cap convex, later depressed. Gills thin, gray-white, decurrent. Poisonous, causing sweating and vomiting. Dia. 0.7-1.5 in.; ht. 0.4-1.1 in.

LACCARIAS differ from Clitocybes in having thicker, more separated gills, which are adnate (broadly attached to the stem), decurrent, or have a decurrent tooth. Older gills are often powdered with white spores, which are large, round, and rough.

WAXY LACCARIA, common in woods and open fields, is so variable it may be difficult to identify. When young, the widely spaced gills are flesh-colored. The small caps, sometimes depressed in the center, vary from pale fleshy red to gray or buff. Waxen when wet. Edible but of poor quality. Dia. 1-2 in.; ht. 3-4 in.

AMETHYST LACCARIA is sometimes considered a variety of the Waxy Laccaria. The small cap, often centrally depressed, is brown, violet, or amethyst when moist, becoming purple-gray when dry. The color of the gills fades less rapidly. Amethyst Laccaria grows in damp, shaded places. Edible, but poor. Dia. 1-2 in.; ht. 2-4 in.

PURPLISH-OCHRE LACCARIA, the largest *Laccaria,* has a scaly stalk. Cap purple-brown when moist, grayish when dry. The gills are purplish. Edible. Dia. 2-5 in.; ht. 3-6 in.

COLLYBIAS have a tough or gristly stem, with no ring or volva. In young specimens, the fleshy cap has a rolled-under margin. Spores are white. Gills vary with the species—thin or broad, close or separated, but most are joined to stem or nearly so (adnate or free).

VELVET-FOOTED COLLYBIA has a brown, velvety-haired stem when mature. Cap is thin, sticky, reddish yellow or dark brown. Gills are broad, somewhat separated. Clusters grow on stumps, and during mild winters survive freezing and thawing many times. Not strongly flavored but good to eat. Dia. 1-3 in.; ht. 1.5-2.5 in.

ROOTING COLLYBIA, of open woods, has a flat, grayish-brown or smoky-brown cap that becomes sticky when moist. It usually has an umbo surrounded by radial furrows. Gills are broad, widely spaced, and not quite attached to stem, which has a long, rootlike extension. Edible but often tough. Dia. 1-4 in.; ht. 4-8 in.

BROAD-GILLED COLLYBIA has a thin, gray to blackish-brown cap. Gills are broad and somewhat far apart. Strong rootlike strands of white mycelium at the base of stem are often pulled up when mushroom is picked. Dia. 3-6 in.; ht. 3-5 in.

BUTTERY COLLYBIA is waxy or buttery to the touch when wet. Reddish-brown cap becomes lighter from the center outward as it dries. The tapered stem has a thickened base; gills, almost free, are thin and crowded. Common in woods. Edible but tough. Dia. 2-3 in.; ht. 2-4 in.

TUFTED COLLYBIA has a thin cap that whitens as it dries. Gills are narrow, close, almost free from top of the reddish-brown stem. Grows in clusters. Edible. Dia. 1-2 in.; ht. 2-3 in.

MARASMIUS are mushrooms that are smaller but like Collybias. They shrivel when dry and revive with moisture. Most species are tough, growing on fallen leaves or small twigs. The spores are white.

LITTLE WHEEL MARASMIUS has dull-whitish or light-brown caps with a small depression in the center. The margin of the cap is furrowed. The gills are few, far apart, and attached to a collar that encircles the stem. Grows in clusters. Not usually eaten because they are so small. Dia. 0.5-0.6 in.; ht. 1-1.5 in.

BLACK-STEMMED MARASMIUS sometimes grows in a dense carpet on pine or spruce needles. Gills are few in number, far apart, and attached to the black stems, which are horny, tough, often contorted. Too small to eat. Dia. to 0.4 in.; ht. to 2.5 in.

GARLIC MARASMIUS has a small cap, at first dull red, becoming whitish, wrinkled, and crisp on drying. Stem is reddish and shining. Gills few, attached to stem. Grows on decaying twigs and needles. Edible, with strong flavor and odor of garlic. Dia. 0.5-1 in.; ht. 1-1.5 in.

ACRID MARASMIUS has a yellowish-brown cap, later becoming pale. Brown stem has white, cottony tufts at base. Tough, bitter, peppery, and possibly poisonous. Dia. 1-2 in.; ht. 2-3 in.

FAIRY RING MARASMIUS is white to pale tan or reddish. Hundreds may sprout, forming a loose circle, or fairy ring, as do many other mushrooms. Gills are creamy, far apart, often with smaller gills between. Stem rough, scaly. Edible; taste pleasant and odor fragrant. Dia. 1-2 in.; ht. 1.5-4 in.

CAPPED MYCENA has a striated cap with an umbo. Variable in color from gray to brown. Gills white, connected by veins. Stem silvery gray, smooth, and rooted at the base. Clusters grow on dead leaves and wood. Small but delicately flavored. Dia. 1-2 in.; ht. 2-5 in.

MYCENAS are closely related to Collybias (p. 65), but the cap is small, thin, and conical, with radiating lines or grooves extending to the margin. The margin is straight, never rolled under. The species are mostly small and difficult to identify without magnification. Stem has no ring or cup. Gills are thin, adnate (broadly attached to the stem), and usually pale.

BLEEDING MYCENA has a small cap with toothed margin and white to reddish-brown gills. Stem has a whitish bloom at first, then turns purplish. Breaking stem releases a blood-red juice. Grows in tufts on stumps. Edible. Dia. 1-1.5 in.; ht. 2-3 in.

CLEAN MYCENA has a cap varying from rose to purple; darker when damp. Cap has umbo when young; expands later. Gills white or pink. Stem smooth, shining; hairy or woolly at base. Has radish odor. Possibly poisonous. Dia. 1-3 in.; ht. 1.5-4 in.

MEADOW HYGROPHORUS has buff or yellowish cap with a reddish tint. Umbo when young; flattened later; margin cracked when dry. Gills decurrent, white or yellowish. Found in old pastures and clearings. Edible, with delicate flavor. Dia. 1-3 in.; ht. 2-3 in.

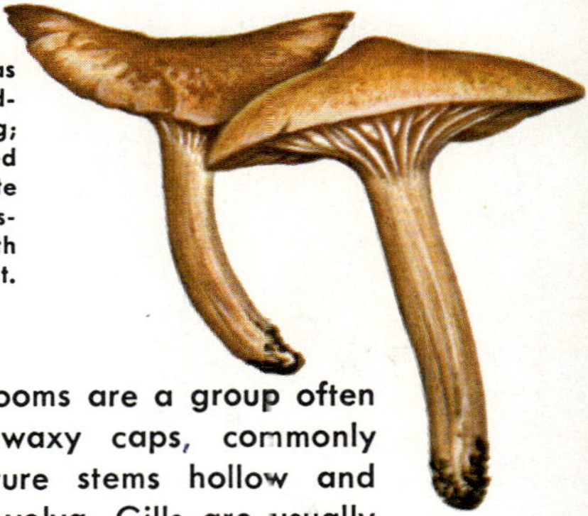

HYGROPHORUS mushrooms are a group often brightly colored, with waxy caps, commonly sticky when moist. Mature stems hollow and twisted, with no ring or volva. Gills are usually thick, relatively far apart, and either adnate or decurrent. Spores are white, rarely colored. Most species are edible but not very desirable.

CONICAL HYGROPHORUS has a conical cap at first, later expanding; sticky when moist, shining when dry. Gills are thin and free from stem, often yellowish. All parts become black when touched. Probably poisonous. Dia. 1-2 in.; ht. 2-3 in.

VERMILION HYGROPHORUS has a bright-red, depressed cap that fades to orange. Cap is dry, not slimy. Gills are yellow to red, and the smooth, shining, slender stem is scarlet. Grows in open woods and swampy ground. Edible. Dia. 0.7 in.; ht. 1-2 in.

ELM TREE PLEUROTUS grows on elms and other hardwoods. Usually off-center, the stem is thickest near the base. The buff cap, at first convex, becomes flat, cracked, and scaly with age. Gills are widely spaced; the spores are white. This mushroom appears in late summer or fall, sometimes growing in clusters, and persists until heavy frosts. Edible when young. Dia. 3-6 in.; ht. 2-4 in.

PLEUROTUS mushrooms grow on logs, stumps, and living trees, often at considerable heights. They never grow on the ground. Stems (occasionally absent) are usually short, lateral, and attached off-center. Gills may be adnexed, as in *Tricholoma* (p. 60), or decurrent, as in *Clitocybe* (p. 62). Spores are usually white, occasionally pink or faint purple. All are fleshy and somewhat tough. Includes a number of flavorful species.

OYSTER PLEUROTUS is white to gray, becoming yellowish with age. It has white spores. Grows in tight clusters, one cap above the other. Gills decurrent; stem, if present, is short. Edible, tasty, toughens with age. Dia. 3-5 in.; ht. 0.2-1 in.

SAPID PLEUROTUS is similar to the above, but the stem is more pronounced and cap is often depressed at the center. Its color is lighter, and the spores are pale lilac. Edible and similar in flavor to the Oyster. Dia. 2-5 in.; ht. 1-2 in.

LENTINUS is a genus similar to both *Pleurotus* and *Panus*, but gill edges are saw-toothed. Spores white. Scaly Lentinus grows in clusters on posts. Scaly cap is fleshy; hard when dry. Edible when young; licorice odor. Dia. 3-4 in.; ht. 1-3 in.

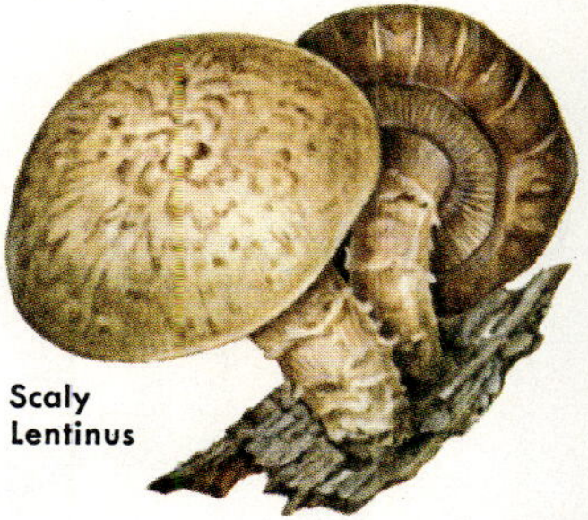

Scaly Lentinus

PANUS mushrooms have smooth-edged gills. Stem, if any, may be central or lateral. Astringent Panus has kidney-shaped caps with scurfy scales. Gills are crowded; spores white. May be phosphorescent. Possibly poisonous. Dia. 0.5-1.5 in.; ht. 0.1-0.3 in.

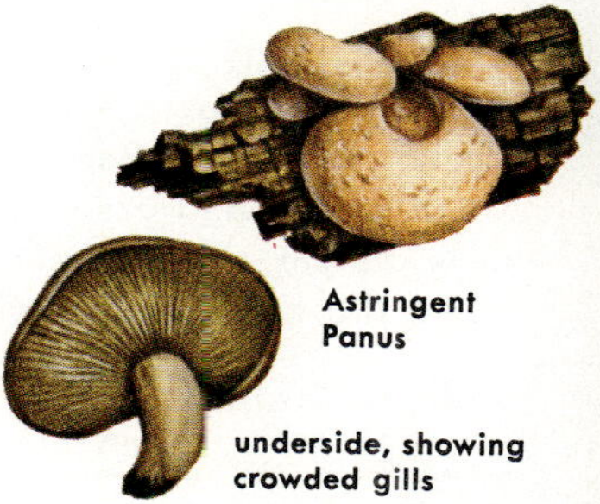

Astringent Panus

underside, showing crowded gills

SCHIZOPHYLLUMS have thin, tough caps with radiating gills that are split along the free edge. Margins of caps are rolled back when dry. Stem may be absent; spores usually white. Common Schizophyllum is the only abundant species. Kidney-shaped caps often lobed, hairy. To 1.5 in. wide.

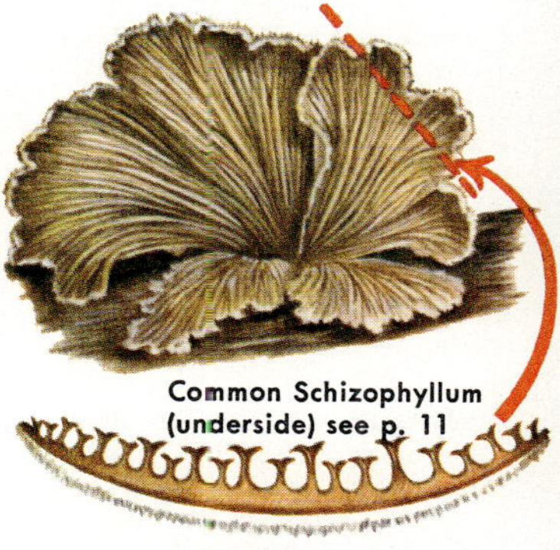

Common Schizophyllum (underside) see p. 11

section through cap, showing split gills

DELICIOUS LACTARIUS is a well-known edible mushroom. Cap may be zoned in shades of orange. Gills and stem are orange to yellow. Greenish stains or blotches appear where plants are bruised. Wounds also turn green. "Milk" is reddish orange. In coniferous and hardwood forests. Dia. 2-5 in.; ht. 1-4 in.

LACTARIUS is a group of large, firmly fleshed fungi without a ring or volva. All exude a white or colored "milk" when cut. Some caps become funnel-shaped when mature and are zoned with concentric markings. Gills adnate or decurrent; spores white or yellowish. Some are edible, many acrid, others poisonous.

SWEETISH LACTARIUS is common and variable. It is tawny or brownish red. The close, narrow gills are paler, sometimes white. All of the many varieties are edible, but some may be bitter. Dia. 0.5-2.5 in.; ht. 1-3 in.

BROWN VELVET LACTARIUS has a dry, dark-brown, velvety cap, flat or slightly depressed. White or yellowish gills slowly become pink when bruised. "Milk" white. Edibility doubtful. Dia. 1-4 in.; ht. 2-4 in.

WOOLLY LACTARIUS usually has a zoned, pinkish cap with depressed center and shaggy margin. Gills thin, narrow, pinkish, and decurrent. "Milk" white and acrid. Probably poisonous. Dia. 2-4 in.; ht. 1-2.5 in.

ORANGE-BROWN LACTARIUS has a tawny cap. White to yellowish gills turn brown when bruised. White "milk" flows copiously. Odor unpleasant when drying. Edible and esteemed by some. Dia. 2-5 in.; ht. 1-4 in.

INDIGO LACTARIUS is strikingly colored. Cap, depressed or funnel-shaped, is indigo with a silvery luster, becoming grayer when old. Gills, stem, and "milk" are dark blue. Edible. Dia. 2-6 in.; ht. 1-2.5 in.

COMMON LACTARIUS has a yellow or tan cap, becoming funnel-shaped with age. Stem paler and slippery. Gills and "milk" white to yellowish. Bruise greenish. Doubtful edibility. Dia. 3-6 in.; ht. 1.5-4 in.

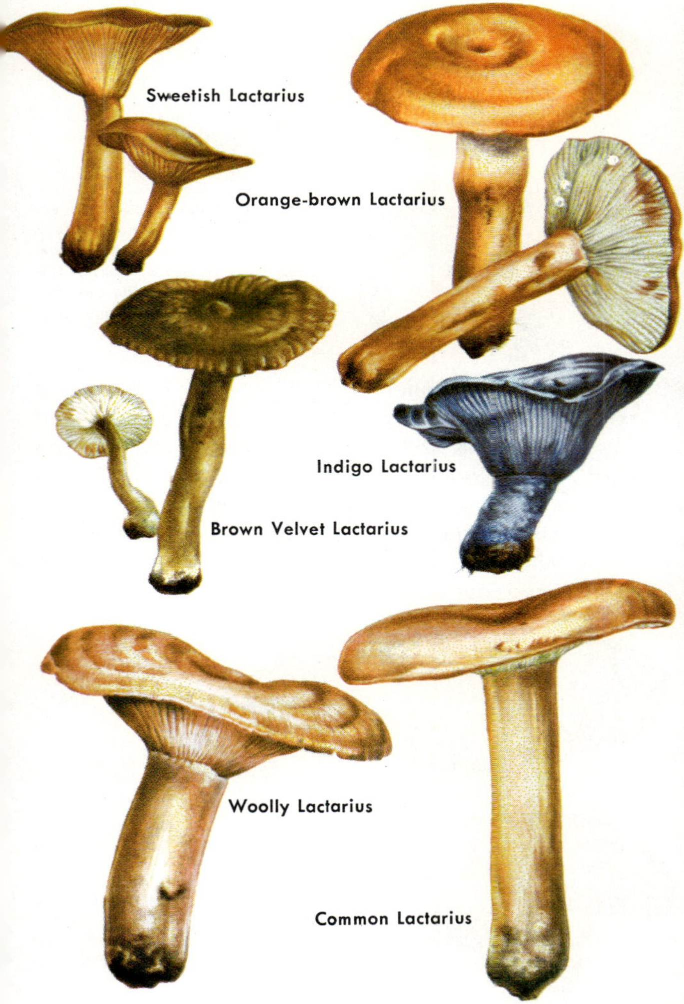
Sweetish Lactarius
Orange-brown Lactarius
Indigo Lactarius
Brown Velvet Lactarius
Woolly Lactarius
Common Lactarius

EMETIC RUSSULA has a convex, flat, or depressed cap. It is rosy or blood red; paler or yellow when old. It peels easily. The broad, nearly free gills and the stem are white. Poisonous. Dia. 2-3.5 in.; ht. 1-3 in.

RUSSULAS are similar to *Lactarius* (p. 72) but never exude "milk." Gills are usually adnate. Both cap and gills are dry and brittle. The stout stem lacks ring and volva. Cap colors commonly change with age. Spores white or yellowish, rarely green. Found in open woods and clearings. The species are difficult to identify. Those with a mild taste are edible; those that are acrid may be poisonous but are not known to be lethal.

FETID RUSSULA is acrid and has a strong, fetid odor. The cap is slimy, its margin striated and with tubercles. Gills are white or straw-colored, often dingy when bruised. Studded with red droplets when young. Probably poisonous. Dia. 3-5 in.; ht. 2-3 in.

BLACKENING RUSSULA has a gray, rounded, depressed cap that turns brown or black. Slimy when moist. White flesh becomes reddish then black when bruised. Edible, with a mild taste. Old caps may be parasitized by small gill fungi. Dia. 3-6 in.; ht. 2 in.

ENCRUSTED RUSSULA grows in woods and open places. The cap —sometimes tan but often tinged with yellow-green—is marked with small, crustlike scales. Margin is furrowed when mature. Gills and stem are white. Edible. Dia. 3-5 in.; ht. 1-2.5 in.

YELLOW RUSSULA, slimy when wet, has a yellow cap, becoming depressed and grooved with age; broad white gills; white to gray stem. Acrid; possibly poisonous. Dia. 2-4 in.; ht. 1.5-3 in.

GREEN RUSSULA has a convex, flat, or depressed green cap with small, woolly or warty patches. Gills are white and crowded, some forked. Edible. Dia. 2-4 in.; ht. 1-2 in.

BIRCH LENZITES forms hairy, sometimes zoned brackets on birches and other broad-leaf trees. Gills are dingy white, often branched. To 4 in. wide.

BROWN LENZITES becomes zoned with dark brown and lighter shades. Hairy brackets sometimes narrowed at base. Destroys conifers. To 3 in. wide.

LENZITES are tough, usually shelflike mushrooms. Because the gills may be joined to form porelike chambers, the genus is considered a link between gill and pore fungi. Spores are white. Too corky to be edible.

CHANTERELLES have branching, dull-edged, shallow gills that, in many species, form veinlike folds down the stem and give the fleshy cap a funnel shape. The spores are white. Includes favorite eating species.

CHANTERELLE, all egg yellow, smells like an apricot. Its depressed cap is often lobed. Edible; common in European markets. Dia. 1.5-4 in.; ht. 1.5-2 in.

VERMILION CHANTERELLE, often abundant, has bright-red cap, gills, and stem. The flesh is white, somewhat acrid. Edible. Dia. 0.5-1.5 in.; ht. 0.7-1.5 in.

PLUTEUS, a pink-spored genus, has neither ring nor volva, and gills are free from the stem. Fawn-colored Pluteus (above) is perhaps misnamed, as cap is more often gray than brown. Edible. Dia. 2-4 in.; ht. 2-6 in.

VOLVARIAS are pink-spored and have a volva but no ring. Caps commonly develop an umbo with age. Both poisonous and edible species are known. *V. esculenta* is the cultivated mushroom of the Philippines.

SILKY VOLVARIA has a silky, shaggy white cap. Gills turn rosy; may have ragged edge. On rotted wood and in humus. Edible. Dia. to 8 in.; ht. to 8 in.

HANDSOME VOLVARIA, bell-shaped at first, becomes flat, with umbo. Sticky cap is white, gray, or brown. Probably poisonous. Dia. 2-4 in.; ht. 4-8 in.

Livid Entoloma

ENTOLOMAS are similar to Tricholomas (p. 60) but have pink spores. Gills white, then pinkish. Margins of fleshy caps at first incurved. Many species poisonous or suspect. Livid Entoloma has a smooth cap with a wavy edge. White flesh first smells like cucumbers, later has a foul odor. Poisonous. Dia. 2-5 in.; ht. 3-5 in.

LEPTONIAS resemble Collybias (p. 65) but have pink spores. Caps thin, margins at first incurved. Gills loosely attached to stem. Most are small. Steel-blue Leptonia grows in pastures. Cap is violet to steel blue; gills light blue. Dia. 1-2 in.; ht. 2-3 in.

CLITOPILUS a genus like *Clitocybe* (p. 62) but with rosy spores. Gills somewhat decurrent. Abortive Clitopilus has a dry cap, at first covered with silky hairs and later smooth. Caps and stems not always developed. Edible. Dia. 2-4 in.; ht. 1.5-3 in.

Steel-blue Leptonia

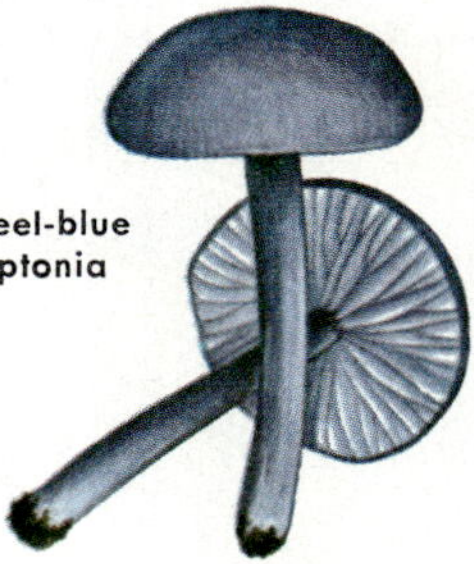

Abortive Clitopilus

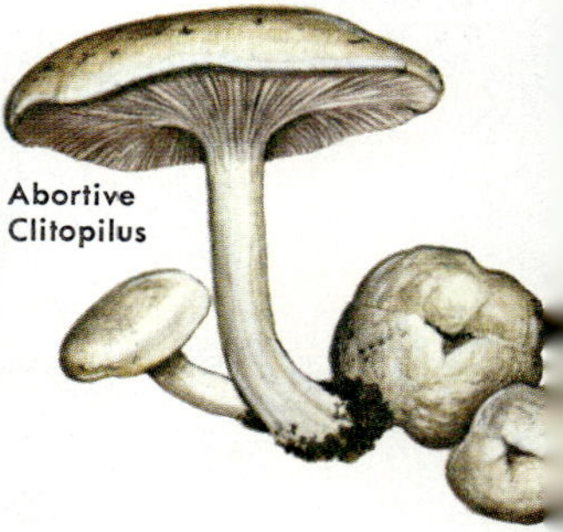

CORTINARIUS is a large genus in which the species have cinnamon-brown or rusty-brown spores. In young stages, a cobwebby veil connects the cap and stem. Gills adnate and powdery with spores. The edible Violet Cortinarius has a dark-violet cap. Violet gills rust-colored when mature. Dia. 2-4 in.; ht. 3-5 in.

Violet Cortinarius

INOCYBES usually have scaly or silky caps. Spores yellowish-clay color. Gills never powdery, free to decurrent. Many species are poisonous. Earth-leaf Inocybe has an umbo when mature. All of the many varieties are poisonous. Dia. 0.8-1 in.; ht. 1-2 in.

NAUCORIAS are small brown mushrooms that resemble *Collybia* (p. 65) but have rust-brown spores. Few species have been tested for edibility. Common Naucoria (edible) grows in grassy spots. Shiny stem is white, pithy. Dia. 0.5-1.5 in.; ht. 1-2.5 in.

Earth-leaf Inocybe (Lilac variety)

Common Naucoria

YELLOW PHOLIOTA (also *Togaria* or *Phaeolepiota*) has somewhat powdery caps to which remnants of the veil may stick. Widespread, but in N.A., most common on Pacific Coast. Edible. Dia. 7-11 in.; ht. 8-12 in.

GYPSY PHOLIOTA has a dry, silky-smooth cap with wrinkled margin. At the stem's base there is a hint of a volva, a reason for sometimes placing this species in a separate genus. Edible. Dia. 2-4 in.; ht. 2-5 in.

PHOLIOTAS are similar to Armillarias (p. 56). Stems usually have a ring; gills adnate. Spores are yellow-brown. *P. autumnalis* (not shown) is poisonous.

SHARP-SCALE PHOLIOTA has spikelike scales over its cap, which is sticky when moist. This edible species grows in clusters on stumps and logs in northern forests. Dia. 1-3 in.; ht. 2-4 in.

EARLY PHOLIOTA appears in spring and early summer in woods and pastures. Its white, yellow-tinged cap smooth; ring on stem soon disappears. Edible. Dia. 1-2 in.; ht 1.5-3 in.

Commercial mushrooms are grown in trays in caves or cellars.

AGARICUS mushrooms resemble Lepiotas (p. 57) but have purple-brown spores. Gills are usually white at first, then pink, and finally purple-brown. Stem and cap are easily separated.

MEADOW MUSHROOM grows in pastures and meadows throughout U.S. and Europe. Cap is dry and white in button stage but becomes brown as spores mature. Some are covered with brown scales. Gills pink, becoming chocolate brown. Ring delicate. Varieties of the prized and edible Meadow Mushroom are cultivated commercially. Dia. 1.5-4 in.; ht. 2-3 in.

HORSE MUSHROOM, an edible *Agaricus* of fields, pastures, and lawns, has an anise odor and sweet taste. Gills, at first white, turn dark brown. Hollow stem has a thick ring, with its lower surface frequently split. Dia. 3-8 in.; ht. 2-8 in.

BLEEDING AGARICUS is brown-capped with darker scales. Bell-shaped cap flattens as plant matures but may retain an umbo. Bruised stem and flesh become red. Stem base may be bulbous. Dia. 2-4 in.; ht. 2-4 in.

SYLVAN AGARICUS, the Forest Mushroom, has a silky, white to yellowish cap; gills pink at first, later darkening. Stem base bulbous. Edible but may be confused with Flat-capped Agaricus below. Dia. 2-4 in.; ht. 3-5 in.

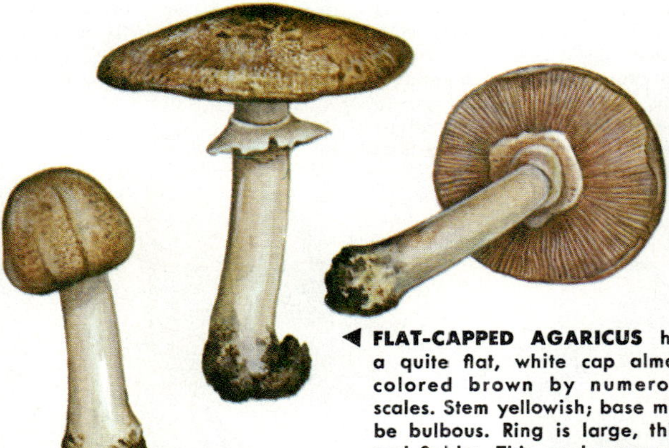

◀ **FLAT-CAPPED AGARICUS** has a quite flat, white cap almost colored brown by numerous scales. Stem yellowish; base may be bulbous. Ring is large, thin, and flabby. This mushroom may be poisonous to some people. Dia. 1-4 in.: ht. 3-5 in.

STROPHARIAS are similar to *Agaricus,* but gills are never free. In many, cap is sticky; ring usually persists. Verdigris Stropharia has an azure to blue-green cap, with pigment in a slimy covering. As slime washes off, cap yellows. May be poisonous. Dia. 1-4 in.; ht. 1-3 in.

Verdigris Stropharia

HYPHOLOMAS are similar to Tricholomas (p. 60) but have purple-brown spores. Margin of fleshy cap is incurved at first. Fibrous stem has a ring represented only by a circle of dark-colored hairs or fibers. Gills may be adnate or adnexed. Although some Hypholomas may be edible, others are bitter and may be poisonous.

BRICK-RED HYPHOLOMA grows in clusters on stumps and logs in late fall. The brick-red caps have white gills when young, dark brown when old. Once believed poisonous, now listed as edible. Dia. 2-4 in.; ht. 3-5 in.

TUFTED YELLOW HYPHOLOMA grows in large clusters. Cap, gills, and flesh yellow. Gills turn green when mature. Common all year where climate permits. It appears to be poisonous to some people. Dia. 1-2 in.; ht. 2-5 in.

SHAGGY MANE is one of the common edible fungi, but must be picked and cooked before the gills darken. The oblong or cylindrical cap has scattered, fluffy scales. Dia. 1.5-3 in.; ht. 3-6 in.

COPRINUS is a genus of black-spored fungi, called Ink-caps. The gills of mature fungi undergo a self-digestion, during which an inky fluid forms, beginning at the margin of the cap and continuing inward to the stem. The entire cap may liquefy in only a few hours. For cooking, use young, freshly picked specimens.

COMMON INK-CAP has a gray, slightly scaly cap. It grows in dense clusters. Edible but may cause illness if alcohol is consumed at the meal or later. Dia. 1-3 in.; ht. 2-4 in.

GLISTENING INK-CAP has shining veil particles on cap when young and is buff or tawny yellow. It grows in clusters at base of trees or stumps. Edible. Dia. 1-2 in.; ht. 1-3 in.

PSATHYRELLAS have caps with fine grooves and straight margins. Spores are brownish to black. Common Psathyrella has an oval or bell-shaped cap. Grows in dense clusters. Edible but small. Dia. about 0.5 in.; ht. 1-1.7 in.

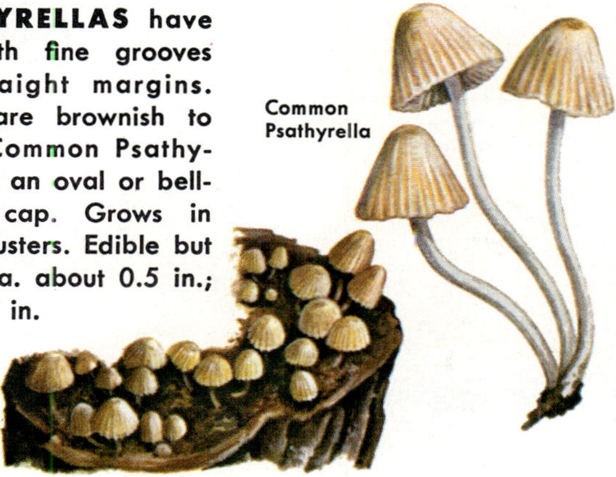

Common Psathyrella

PANAEOLUS is a genus of mushrooms with black spores and a smooth, bell-shaped or conical cap. Gills mottled when young, black later. Common on manure.

BELL-SHAPED PANAEOLUS grows on dung or in rich soil. Reddish-brown caps turn pale and shining when dry. Flesh is red. Gills adnate or adnexed. Poisonous. Dia. 1 in.; ht. 4-6 in.

BUTTERFLY PANAEOLUS is grayish when wet, clay-colored and shining when dry. Cap is hemispherical or has an umbo. Stem is pale; gills broad. Poisonous. Dia. 0.5-1.5 in.; ht. 3-5 in.

GIANT PUFFBALL is one of the largest of all fungi (wt., 10 lbs.; dia., 10-20 in., occasionally larger). Grows in fields, where it is neither abundant nor rare. Edible and fine-tasting while still white and firm.

GASTEROMYCETES

Gasteromycetes include puffballs, earthstars, bird's nest fungi, and stinkhorns. They differ from the pore and gill fungi (p. 46) in that all have a peridium over the spore-bearing surface. The peridium, usually several layers thick, ruptures and exposes the spores when they are ripe and have fallen off the basidia. When they are ready for dispersal, spores may be spread by wind, rain, or animals. Black spore clouds are released from a forcefully ruptured puffball, for example. In some bird's nest fungi, the spore-containing peridioles are literally splashed out of the cup by rain. In some, each sticky peridiole is equipped with a cord that helps to entangle and hold it to a surface until it splits and releases its spores.

Most Gasteromycetes grow in soil; some grow in decaying wood. Some species grow singly, others in clumps. Fruiting seasons vary by species and region, but July, August, and September are usually the most favorable months in north temperate areas.

PUFFBALLS are among the most easily identified Gasteromycetes. They range in size from the round Giant Puffball (opposite) to the 1-3-in. pear-shaped Lycoperdons. The fleshy white interior yellows and becomes brown at maturity. All are edible when white and solid, except species of *Scleroderma* (not shown).

◀ **GEM-STUDDED PUFFBALL,** 1-2.5 in. tall, grows in dense clusters on ground or rotten wood. Its warty white surface darkens as it matures. Rounded top tapers to a stalklike base. Tough skin hard to remove. Dia. 1-2 in.

PEAR-SHAPED PUFFBALL, 1.5-2 in. tall, is covered with tiny "spines" and warts that fall off the rounded top as it matures and becomes smooth. Stem tapers to base. Grows in clumps on decaying wood. Dia. 1-2 in. ▶

◀ **BEAUTIFUL PUFFBALL,** 0.5-2 in. tall, may be spherical or have a stalklike base. Found singly on ground. Coat of tapering "spines" darkens at base and is shed from upper part when mature. Edible. Dia. 0.5-2 in.

SPINY PUFFBALL, 1-2.5 in. tall, grows on soil. "Spines" turn brown at tip and a patterned membrane is revealed when they fall off. Mature mushrooms emit purple-brown spores. Edible when young. Dia. 1-1.5 in. ▶

MANY-SACKED PUFFBALL, about 3 in. high, grows in exposed soil. Yellow interior is divided into many sacs, each containing basidia and spores. Fresh mushrooms exude an inky juice. Poor eating. Dia. 3 in.

STALKED PUFFBALLS show only their globular heads when growing in loose soil or sand. In hard surfaces, both cap and stalk stick above—0.5-1.5 in. Edible when young; bitter when mature. Dia. 0.2-0.5 in.

EARTHSTARS split open as they mature. The thick outer layer divides into pointed segments, producing a starlike effect. This reveals an inner ball with a thin, papery coat. Spores are discharged through mouth openings on the top. Mycelium threads often enmesh debris about the outer segments.

TRIPLEX EARTHSTAR, 1-2 in. high, has a tough outer husk that splits at top when mature to form six to eight petals and a rounded collar. Petals unfurl when moist, revealing inner spore sac. Spores discharge through mouth, encircled by a ridge. Dia. 1-2 in.

CROWNED EARTHSTAR, 1-2 in. high, splits into four to eight strongly recurved petals, often raising the whole plant above the ground. A ridge outlines central mouth area of the inner ball. Grows singly or in small groups in woods. Dia. 1-2 in.

COMMON BIRD'S NEST cup is 0.1-0.3 in. high, with a rounded top. As the cup grows, a hairy, brown membrane breaks and exposes a white membrane beneath. This also ruptures and disappears, revealing spore-filled "eggs." Dia. 0.1-0.3 in.

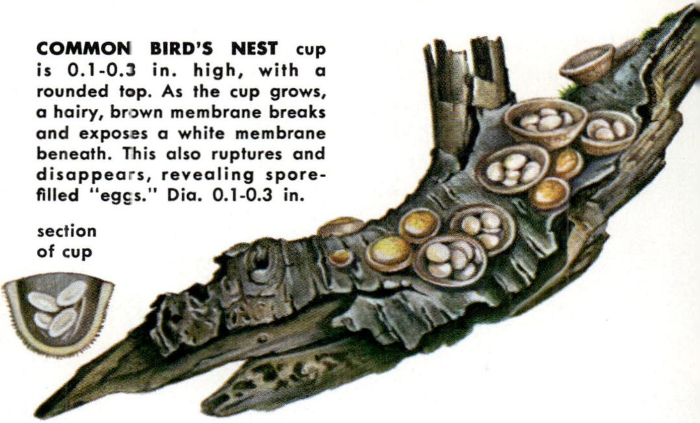

section of cup

BIRD'S NEST fungi are aptly named. These cuplike receptacles, 0.1 to 0.5 in. in diameter, contain several egglike bodies or peridioles. The tiny "eggs" may be attached by a thread to the cup. Within them are spores, released when the covering splits open. Bird's nest fungi grow on twigs, decaying wood, straw, or sawdust.

STRIATE BIRD'S NEST, height to 0.5 in., is cone-shaped, narrowing at base to a stem. Outer surface is hairy. When the fungi are fully opened at maturity, the striated interior and the cord-attached spore balls are revealed. Dia. 0.2-0.4 in.

WHITE BIRD'S NEST may appear singly or in groups. Outer surface of cup white and hairy, darkening with age; inner surface yellowish. Eggs not attached by cords. Cup dia. 0.2 in.

section of cup

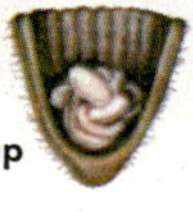

STINKHORNS are egg-shaped and remain underground until the spores mature. Then a stalklike structure emerges from the pouch, or peridium (longitudinal sections, below). The tip of the stalk is covered with a slimy mass of foul-smelling spores. Flies, attracted by the odor, feed on spore mass and spread the spores.

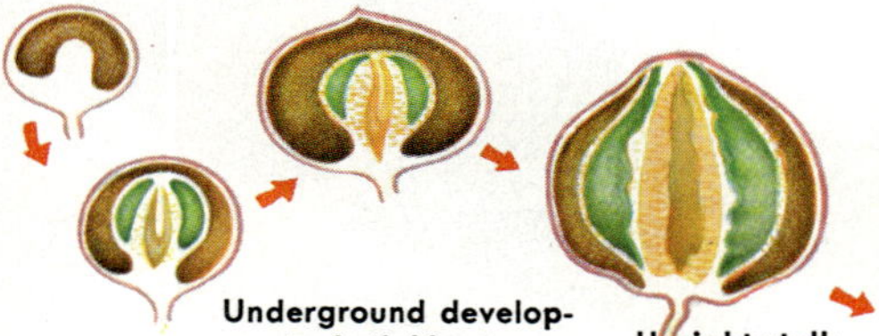

Underground development of stinkhorn

Upright stalk develops in 2-5 hrs.

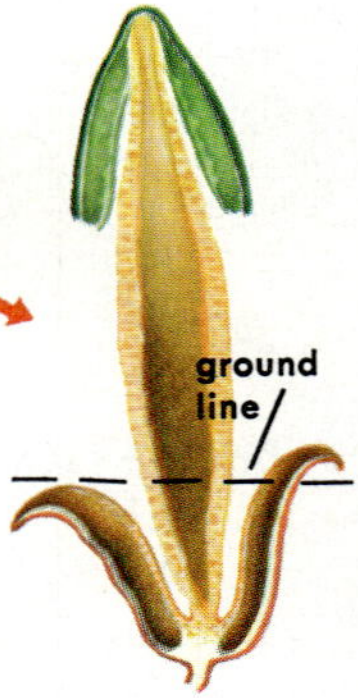

SMALL STINKHORN is considered rare but this may be due to its small size and faint odor. Seldom more than an inch or so tall. Its stalk, divided into three or four columns, is a bright yellow-orange at top.

LATTICE STINKHORN, widely distributed, thrives in warm areas. Open network structure arises from egglike peridium. Red above, paler below; olive-green spores on inner surface. Dia. 5-7 in.; ht. 3-4 in.

COLUMNAR STINKHORN has four or five cinnabar-red, spongy, curved columns that unite at the top. The sticky spore masses are on the inner faces of the columns. Odor fetid. Poisonous. Dia. 2-3 in.; ht. 4-5 in.

Dog Stinkhorn

Common Stinkhorn

Collared Stinkhorn

DOG STINKHORN is rosy red at the top, fading below. The upper part is covered with a dark slime. The hollow, spongy stalk is slender, and there is no distinct cap. The Dog Stinkhorn grows in damp places, on old stumps or rotting wood. Odor is not strong. Ht. 2-3 in.

COMMON STINKHORN may grow on rotting wood or on old sawdust piles. It may be a pest in gardens, as the unpleasant smell of the slimy spores on the cap carries for some distance. Poisonous. Ht. 6-8 in.

COLLARED STINKHORN has a rosy, sometimes white, netlike veil that hangs several inches below the cap. This lacy network makes the fungus more conspicuous after dark and attracts night-flying insects. Ht. 6-7 in.

IMPERFECT FUNGI

Imperfect Fungi include those microscopic fungi whose life cycles, particularly the sexual stages, are not known. They do not form large fruiting bodies but reproduce by various air-borne asexual spores. Molds that mildew walls and spot clothes are included in this group. Many plant diseases and also athlete's foot, ringworm, and other skin infections are caused by these fungi.

Alternaria spores, common in the air, are one cause of hay fever and similar allergies. They are dark-colored and many-celled, with both longitudinal and horizontal divisions. *Alternaria* is responsible for deterioration of paper and textiles, and some species cause plant diseases. Spores of *Helminthosporium* are similar to those of *Alternaria* but are more rounded and divided only horizontally. *Helminthosporium* species infect grasses.

Species of *Fusarium* cause wilts and also deteriorate paper and textiles. Gibberellic acid, used to promote growth in plants, is obtained from *F. monilliforme*.

Other common Imperfect Fungi include *Colletotrichum,* which causes blights and anthracnoses of plants, and *Pestalotia,* cause of leaf spots and fruit rot.

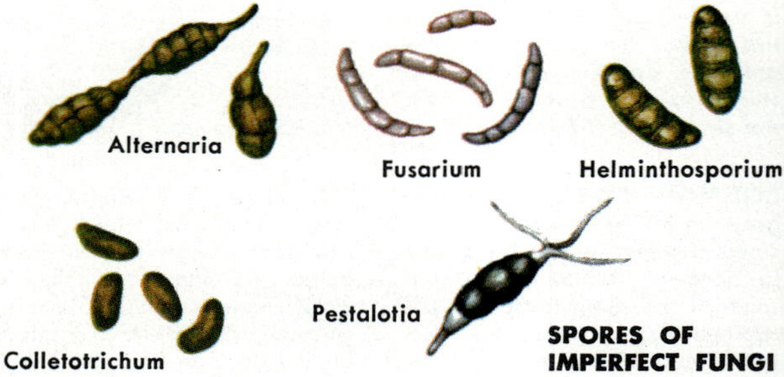

SPORES OF IMPERFECT FUNGI

Blister Lichen, a foliose type common on trees

CROSS-SECTION OF LICHEN THALLUS

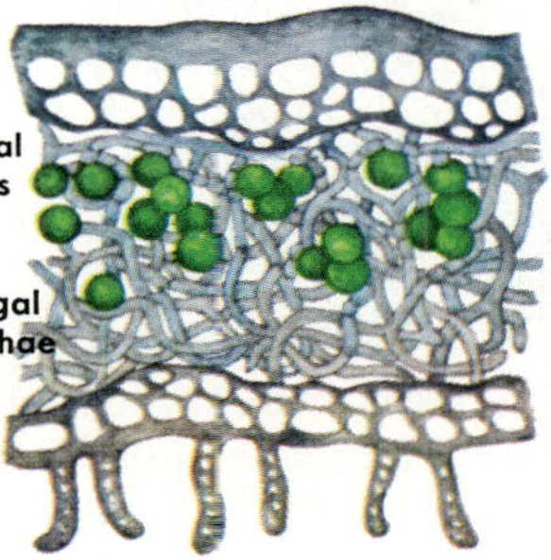

LICHENS

Lichens are pioneer plants which usually grow where other plants do not furnish competition. They are found on rocks, trunks of trees, logs, sand, and bare soil. These slow-growing, long-lived, sun-loving plants flourish in cold, dry climates, in forests and on mountains—wherever the air is clean and unpolluted. Reindeer and caribou eat some lichens; extracts from a few species are used as antibiotics. The best-known use of lichens has been in making dyes for Harris tweeds and for litmus paper.

Lichens are "dual organisms" being composed of two kinds of plants—a fungus and an alga. These seem to live together for mutual benefit (symbiosis). The fungus, usually the most conspicuous portion of the plant, supplies the alga with water and minerals and may prevent it from drying out. The alga manufactures carbohydrates which may also be used by the fungus. A cross section of a lichen (above) reveals a layer of algae (usually single-celled green algae) a short distance below the surface.

LICHEN REPRODUCTION

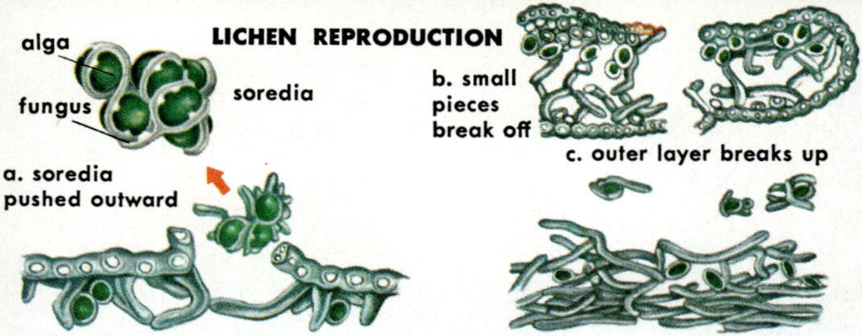

Lichens generally reproduce vegetatively. Some kinds produce dustlike particles (soredia) in which a few algal cells are surrounded by a tangle of fungal hyphae (a, above). These are carried by the wind and may develop into new lichens. Lichens also reproduce when small pieces break off (b) or when the top layer breaks up (c) and bits blow away.

Nearly all fungi in lichens produce spores, but unless the fungal spores are disseminated together with compatible algae, lichens are not formed.

Lichens may be classified as crustose (crusty), foliose (leaflike), and fruticose (shrubby), see p. 95, top, but many intermediate forms exist. A crustose lichen grows flat on rocks or tree trunks and may be embedded in them. The foliose type is attached only in spots; the margins are often lobed and free. The shrubby, fruticose type is a branched plant, upright or hanging.

RED BLANKET LICHEN is an exception to the usual definition of a lichen. This beautiful plant has purple bacteria enmeshed in the fungus rather than an alga. Red Blanket Lichen is a foliose type, common on trees in Florida, the Gulf Coast, and tropical America.

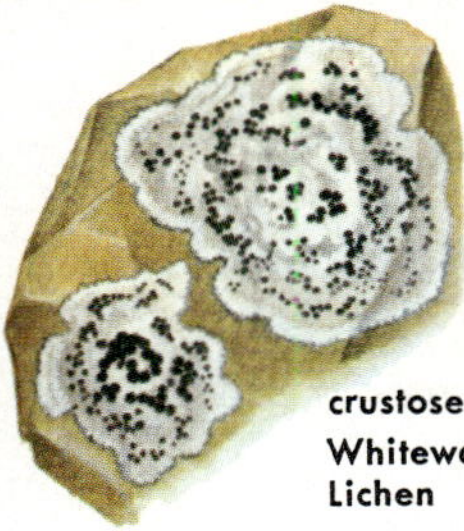

crustose
Whitewash
Lichen

foliose
Dog Lichen

fruticose
Spoon Lichen

CRUSTOSE LICHENS

RED-EYE LICHEN, one of a group called Blood Lichens, grows on bark and sometimes on rocks, from the East Coast to Texas. The crustose thallus (to 1 in. across) is covered with numerous small scarlet disks (fruiting bodies) that often darken and become white-rimmed and broken with age.

PITTED LICHEN is widespread. Its thin, white to gray thallus is partially sunken into the pitted surface of limestone. Its numerous small black fruiting bodies are visible above the thallus. A black pattern, where the fruiting bodies have been embedded, is left in the rock when the lichen finally dies.

CLOT LICHEN, found world-wide in northern regions, forms a cracked and warty crust on evergreens and on rocks at high elevations. Shiny black fruiting bodies, blood red underneath, arise from the crust of united greenish-gray or ash-colored granules. Crust 5 in. across.

CRUSTOSE LICHENS

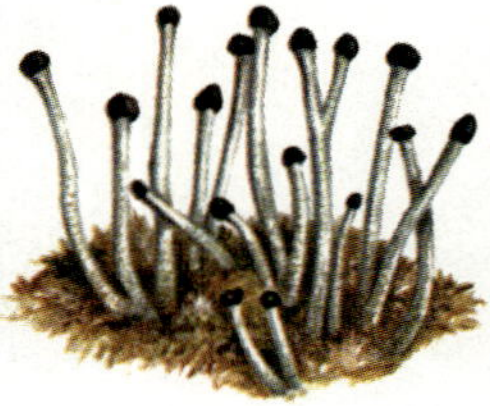

LOLLIPOP LICHEN forms powdery crusts on the ground. Before these disappear with age, short branches, about 0.5 in. high, rise above them. The branches are topped by black fruiting disks. A West Coast species grows to Rockies and in Europe.

MAP LICHEN forms a greenish to bright-yellow, chinky or warty crust on rocks through northern N.A. and southward in the mountains. Numerous small black fruiting disks are partly sunken into the thallus.

PINK EARTH LICHEN is found on sterile soil, as in roadcuts and ditches, in most temperate regions. The gray thallus produces flesh-colored fruiting knobs on 0.2-in. stalks. Cottony fibers fill the knobs.

SCRIPT LICHEN, or Writing Lichen, is cosmopolitan, growing on hardwood trees. Black, straight or curved fruiting disks, protruding above or partly embedded in the crust, superficially resemble handwriting.

ORANGE STAR LICHEN, with its lobed, almost leaflike, yellow-orange thallus (to 4 in. across), has flat, orange fruiting cups. It grows on rocks in north temperate and arctic areas, and atop southern mountains.

FOLIOSE LICHENS

PALE SHIELD LICHEN, found on northern trees and logs, is a leaf-like, grayish-green lichen up to 6 in. across, with pale to dusky-brown, lobed and jagged margins. Fruiting bodies rare. May be confused with the common Boulder Lichens.

BOULDER LICHEN forms 2-in. rosettes on trees and old wood or spreads as tangled mats on rocks. The perforated thallus is black below, with chocolate margins. Several species. In North and in southern mountains.

FAN LICHEN, common on soil, has a 0.5-in., fan-shaped thallus, greenish gray to brown above; white with a network of dark veins below. Brown fruiting disks on lobe margin. The related Dog Lichen (p. 95) is larger.

SHELL LICHEN, found on trees of eastern N.A., has a lead-colored thallus (2-4 in. across), with bluish-black, hairlike strands on the undersurface. Fruiting bodies yellowish brown to dull black.

TOAD SKIN LICHEN, one of the Rock Tripes, has pustules on top of the thallus (sheets often 4 in. across) and pits below. The flat fruiting disks are black. Common on rocks, east of Rockies. Also see smooth species (p. 98).

OREGON LUNG LOBARIA, a foliose lichen, grows on West Coast trees, but its relatives are found throughout N.A. The large, loosely attached thallus is smooth to pitted and ribbed. Variable in color, it is bright green when moist.

SMOOTH ROCK TRIPE has a 2-12-in. oval, brown, foliose thallus, black underneath. A single central cord attaches it to the rocks. Grows throughout the mountains of eastern N.A. Edible (when boiled) as a survival food.

FRUTICOSE LICHENS

HORSEHAIR LICHEN, a beard-like lichen, grows on trees and dead wood through most of N.A. The slender, brown to olive stalks fork repeatedly. On mountains, it also grows on the ground, forming smooth straggly 4-in. tufts.

GOLDEN LICHEN has a tufted, erect or spreading beardlike thallus, yellow-orange to ashy. Many entangled branches. Grows on trees in northern coastal areas. Length 2 in. or more.

BEARD LICHEN is shrubby, tufted, greenish gray or darker. Old Man's Beard, a related form, is more flexible and pendulous. Both are common through much of N.A., growing on trees and old wood, occasionally on rocks. Tufts 2 in. or longer.

FRUTICOSE LICHENS

BRITISH SOLDIERS, or Red Crest Lichen, is one of a large common genus. Flattened thallus scales are greenish gray; branched stalks (to 1 in.) have globular red fruiting tips. Found on soil or decayed wood throughout eastern N.A.

PYXIE CUP, or Goblet Lichen, is common on rocks, soil, and rotten wood. This lobed, scaly thallus has stalked, upright cups (to 1 in.). Branches may develop from cup lip and brown fruiting disks grow on them or on lip. Many varieties in N.A.

SPOON LICHEN has cylindrical or trumpet-shaped stalks (to 3 in.) with small, pale- or dark-brown fruiting bodies on toothed margins. Found at high elevations through much of northern N.A. Usually grows on soil, sometimes on rotted wood.

LADDER LICHEN, with cups arising out of cups, forms tiered stalks to 3 in. high. Upper cups grow from the center of those below, unlike the Spoon Lichen, whose tiered cups ascend from the toothed margins. A common soil lichen of N.A. and Europe.

REINDEER LICHEN is a shrubby lichen that forms patches on soil, often 10 in. in dia. Brown fruiting disks are rare. Found widely in northern and arctic areas; a basic food of reindeer.

EMBRYOPHYTES

Embryophytes are plants that form embryos during one stage of their reproductive process. Most have more complex structures than the Thallophytes. Two main groups of Embryophytes are generally recognized: the Bryophytes, with about 25,000 species, and the Tracheophytes, numbering about 10,000 non-flowering and 200,000 flowering species.

The key to the separation into two divisions is vascular tissue—specialized cells and tubes for the movement of liquids and food. Xylem is the water-conducting tissue and phloem conveys food material. Those plants without vascular tissue or roots, such as mosses, liverworts and hornworts, are Bryophytes. Those with vascular tissue—ferns, fern allies, and all of the seed plants, both cone-bearing and flowering—are Tracheophytes.

Both the sex organs (archegonia and antheridia) and the spore cases (sporangia) are always many-celled in Embryophytes. A jacket of sterile cells surrounds the sex cells, or gametes, and the spores. And when two unequal gametes (the egg and sperm) unite, forming a zygote, this develops into an embryo, an immature growing plant, inside the female sex organ.

Of the Embryophytes, only the cycads, conifers, and flowering plants produce seeds, as did seed ferns, now extinct. The other plants of this sub-kingdom reproduce by non-motile spores, which are blown about by the wind. All Embryophytes have a two-phase life cycle, alternating a generation that produces gametes (sexual) with one that produces spores (asexual). The spore-producing generation (sporophyte) always begins its development as an embryo which is for a while a

DEVELOPMENT OF EMBRYOPHYTES

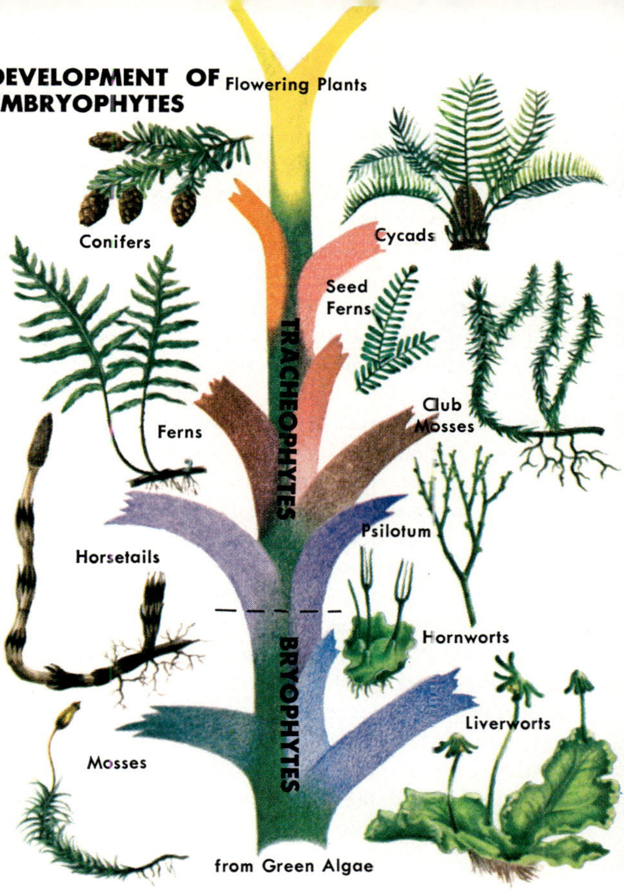

parasite on the gametophyte, or sexual generation. (See pp. 102-103, 117.)

Embryophytes probably developed from the Thallophytes by way of the green algae. Relationships are shown above, a variation of the family tree on pp. 6-7.

MOSSES

Mosses are Bryophytes that grow on soil, rocks, or bark of trees. Some are truly aquatic, living in bogs or even submerged in streams. Mosses grow the world over from the Arctic to the tropics, from mountaintops to seashores. Some prefer acid, others alkaline soils. All lack roots and true stems; the leaves do not have veins as do the leaves of vascular plants. Most are just a few inches high, though some are much smaller and a few aquatic species may be up to two feet long.

Mosses grow from spores produced in capsules. When a spore lands on a spot with sufficient moisture, the walls swell and a filamentous green thread (protonema) grows out. This branches and, in growth, resembles a green alga. Eventually the protonema produces buds, which grow into the leafy stalks that most people recognize as moss plants. The protonema usually dies and the leafy stalks develop tiny rootlike structures at the base and continue to grow as independent plants.

These leafy plants are the gametophytes; usually one plant is male and another female, although some species produce both sex organs on the same plant. The upper leaves of the male plant may form a flat rosette around the male sex organ (antheridium). Sperm cells are produced and these must swim through a film of water formed by rain or dew to the top of female plants where the female sex organ (archegonium) is located.

The sperm unites with the egg to form a zygote. This divides many times and grows to form a sporophyte, usually a long stalk with a capsule on top. The sporophyte foot remains embedded in the gametophyte, which becomes host to the parasitic spore-producing plant.

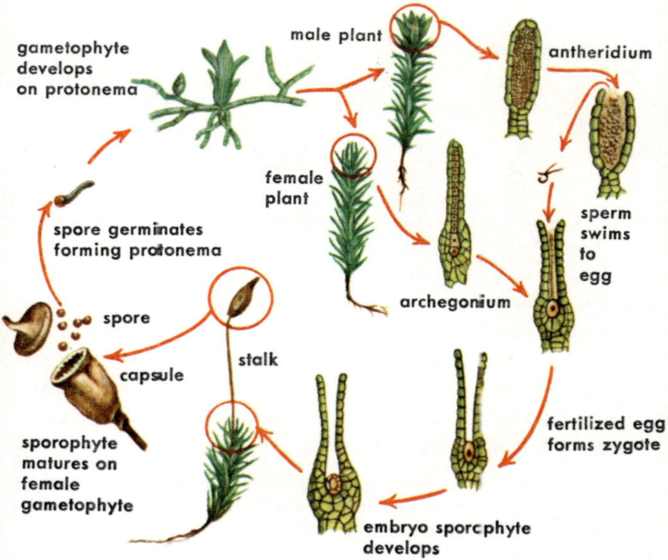

LIFE CYCLE OF A MOSS

Within the capsule numerous spores develop. These escape, germinate, grow into protonema, and repeat the cycle (above).

Many mosses are difficult to identify without a hand lens. The sporophyte capsule is complex. For identification note the capsule with its lid, which falls off when the spores mature (p. 104). Beneath the lid and around the capsule opening is a single or a double row of teeth (peristome). The number and form of the teeth are important identifying characteristics. Other features used are size, shape, and position of the capsule; length of stalk (seta) supporting the capsule; position and structure of leaves.

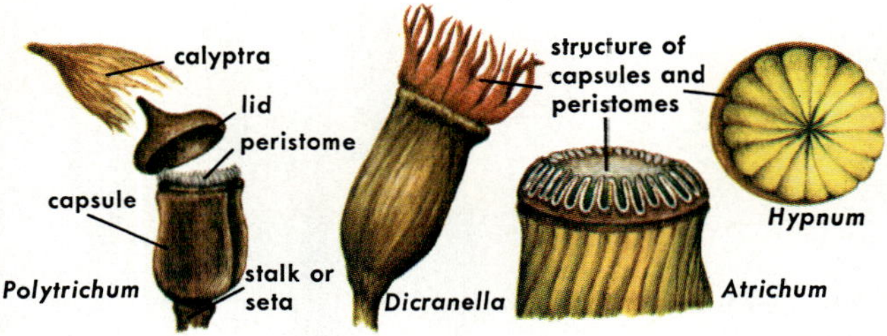

Mosses are of more indirect than direct benefit to man. Sphagnums are mixed with soil to retain moisture and are used in seedbeds where their additional antibiotic action aids young seedlings. In bogs, mosses decay to form peat, a major fuel in northern regions. They also help prevent erosion and flooding by holding moisture from rains and melting snows.

SIZE OF MOSSES is indicated in the captions, usually at the end. The height (Ht.) of the gametophyte is given first (though the gametophyte may be flat). This is followed by the length (L.) of the stalk and the capsule of sporophyte.

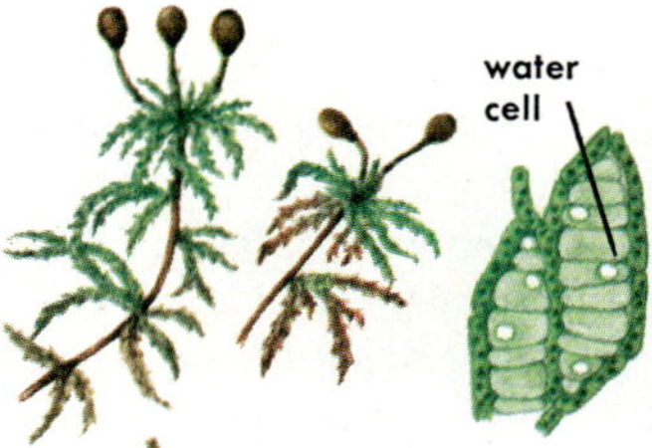

SPOON-LEAVED SPHAGNUM, is one of 300 or more species of peat mosses of cool temperate regions. All peat mosses have two kinds of leaf cells: small and green for food-making and large, dead, water-storage cells (left). Sporophytes are rare. *Sphagnum* grows in bogs and shallow lakes, often forming floating islands. Entire plant may rise 1 ft. tall.

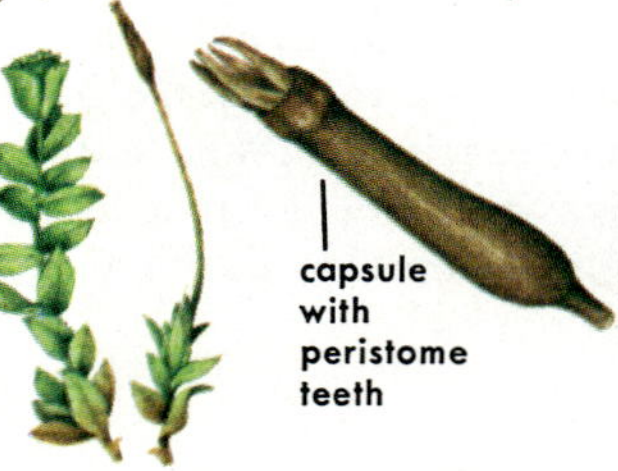

FOUR-TOOTH MOSS is frequently found on decayed stumps and old logs in northern N.A., Europe, and Asia. Some of the leaves form cuplike clusters. The four long teeth of the peristome can be seen without a lens. Ht. 0.5 in.; L. 1 in.

ELF-CAP MOSS, found on moist soil or decayed wood, has a bug-like capsule perched on a stout red stalk. The leaves disappear long before the capsule's green autumn coat turns to reddish brown or the top of the capsule flattens. This prostrate moss is almost microscopic, except capsule and stalk (both to 0.5 in.).

HAIRCAP MOSS, a cosmopolitan species, grows on damp ground. The capsules are four-angled; the stalk long, the calyptra hairy. The leaves, widespread when wet, are long, narrow, and toothed on the margin. This species is often illustrated as a typical moss, but is no more common than many other species. Ht. 6 in.; L. 4-6 in.

NUT MOSS plants crowd together, forming dark-green mats on moist banks. Hair-tipped leaves surround the tilted, nut-like, almost stalkless capsules. Ht. 0.1-0.2 in.; capsule 0.2-0.3 in.

SPINELEAF MOSS is common on shady banks. Wavy leaves have minute teeth and a pronounced midrib. Capsules strongly curved. Plants from the Rocky Mountains westward have larger leaves. Ht. 1-2 in.; L. 1-2 in.

FALSE HAIRCAP MOSS, found on roadside banks, grows on a green mass of persistent protonema. Leaves few, narrow, and sharp-pointed. Capsule has gray calyptra. Ht. 0.3 in.; L. 1 in.

SILKY FORK MOSS forms dense patches on wooded banks. Upright plants are often branched; leaves smooth with long, tapering tip. Capsule has a beaked lid. Peristome teeth forked. Ht. 0.5-2 in.; L. 0.2-0.5 in.

BURNED GROUND MOSS is common on walls, sidewalks, and dry ground. Leaves have rolled-back margins except at toothed tip. Furrowed capsule has conical lid. Ht. 0.5-0.7 in.; L. 0.6 in.

BROOM MOSS forms light-green tufts in woodlands. Stems are tall, with tapering leaves normally bent in same direction, appearing wind-blown. Cosmopolitan in north temperate areas. Ht. 2-4 in.; L. 1-1.5 in.

WAVY BROOM MOSS has long, wavy leaves with a silky luster. These merge into robust, bright, glossy mats. Capsules are clustered, each on a long stalk. Grows in shade on soil or stones. Ht. 2-5 in.; L. 1-1.5 in.

WHITE CUSHION MOSS grows in poor, acid soil, often producing spongy, dense cushions several inches deep. Plant is dirty white, becoming blue-green when wet. Sporophytes seldom formed. Ht. to 4 in.; L. 0.5-1 in.

CORD MOSS frequently grows on burned or limed soil or in limestone areas. The closely overlapping leaves form a bud-like head. The mouth is on the side of the pear-shaped capsule. Peristome teeth have crossbars. Ht. 0.2-0.5 in.; L. 1-2 in.

URN MOSS appears frequently on shaded roadsides, in open woods, and on damp soil in greenhouses. Urn-shaped capsules have erect cap and no peristome; mature in May. Ht. 0.1-0.5 in.; L. 0.2-0.5 in.

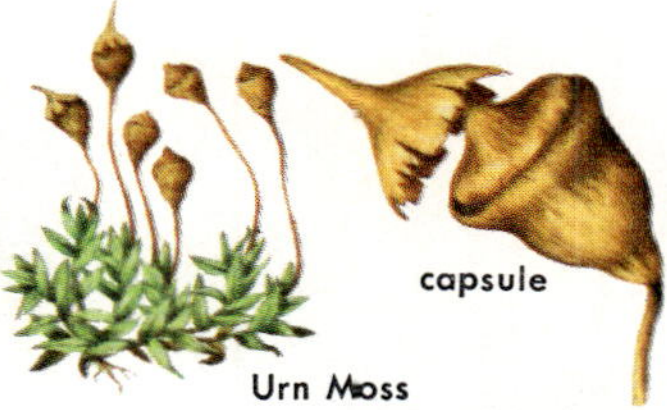

APPLE MOSS has small, apple-shaped capsules, furrowed when dry. Often makes woolly patches on rocky soil of woodlands. Leaf margins have double row of tiny teeth. Ht. 1-3 in.; L. 0.5-1 in.

Apple Moss

INDIAN BRAVE MOSS, of northern regions, has an erect calyptra jutting from behind the bent capsule like a single feather of a warrior. Orange leaf base is noticeable where it overlaps stem. Ht. 2-4 in.; L. 0.7-2 in.

Indian Brave Moss

ROSE MOSS has creeping stems with upright branches. Lower leaves scalelike; upper leaves large, forming rosettes. Fruiting uncommon, but sporophytes occasionally found. May form mats. Ht. 1.5-2 in.; L. to 2 in.

Rose Moss

SILVER MOSS is found on waste ground, walls, roofs, and even between paving stones. The overlapping leaves lose their chlorophyll with age, creating mats of tiny silvery shoots. Capsules droop. Ht. 0.2-0.5 in.; L. 0.5 in.

TWISTED MOSS grows in dense tufts on limestone. Peristome teeth are twisted, as are the long, narrow leaves, which have a green V near their translucent base. Ht. 0.5-3 in.; L. 0.5-2 in.

WALL MOSS is common on stone walls. Leaf tip has a hair point. Dry leaves are twisted, as are the peristome teeth. Many similar species are found in N.A. Ht. 0.2-0.5 in.; L. 0.5-1 in.

WHITE-TIPPED MOSS is a branched and spreading plant (to 4 in. long) with colorless tips on upper green leaves. Common on rocks and stone walls. Leaves may hide short-stalked capsules. Plant flat; L. to 0.2 in.

▼

FRINGE MOSS, found on wet rocks along streams, has long (to 4 in.), prostrate stems with many erect branches. Broad leaves entire or toothed at apex. Capsules red-brown; lid long-beaked. Ht. 1-2 in.; L. 0-2-0.5 in.

▼

enlarged

LUMINOUS MOSS is found at cave entrances and in dark areas in the woods. A cobweb of fine green protonema and a few tiny plants produce a golden-green glow by reflected light. Ht. 0.5 in.; L. 0.2-0.5 in.

SHAGGY MOSS is a large plant often found on logs or soil in moist woods. Stiff stems are irregularly branched. The double-ribbed, serrated leaves stand out at right angles from the stem. Ht. to 6 in.; L. 1-1.5 in.

STAR MOSS is common in N.A., Europe, and Asia, growing on moist soil or rotten wood. Leaves form rosettes at tip of male stem. Sporophytes arise in spring and remain until fall. Ht. 1 in.; L. 1-1.5 in.

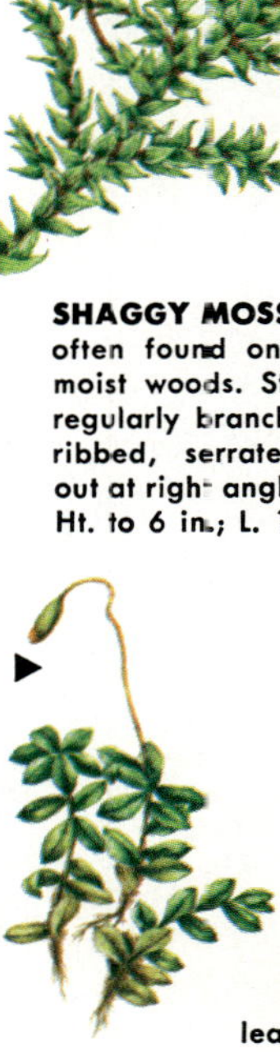

FOREST-STAR MOSS has long, narrow leaves with paired, marginal teeth. Midrib stops before end of leaf. Capsules on gracefully curved stalks. Found on soil and rocks in shady, humid woods of eastern N.A. Ht. to 3 in.; L. 1-1.5 in.

SHINY MOSS, or Common Hypnum, is found on decaying wood and in peaty soils east of Rockies and in Europe. Leaves shiny, straight, and untoothed, forming flat sprays (to 3 in. long) in wide mats. Slender capsules, often numerous, have a short-beaked lid. Usually flat; L. 0.5-1 in.

MOUNTAIN FERN MOSS forms loose mats in cool, moist mountainous regions. Leaves are larger and more easily seen on main stems. Each year's new shoot rises from the middle of the previous year's main stem. Ht. flat to 2 in.; L. 1 in.

TREE MOSS has creeping stems with upright branching stems that give the appearance of tiny trees. Spreading leaves toothed near tip. Grows in clumps or as separate plants in swampy ground in mountainous regions of northern N.A. and in Europe. Ht. to 4 in.; L. 1-2 in.

SLENDER MOSS is flattened in growth habit; branches arise irregularly. The glossy leaves, also flattened, are toothed at the tip. They appear to form two opposing rows. The lid of the capsule has a short beak. Ht. flat to 0.3 in.; L. 1-2 in.

OSTRICH PLUME MOSS, the most beautiful of the feathery mosses, grows on old logs, often covering them completely. Branches are regular; leaves without midribs, curved and soft. It is a light yellowish green in color. Capsules long and curved. Ht. flat to 4 in.; L. 0.5-1 in.

FEATHER MOSS grows in loose patches, high on trunks of trees, in moist woods of temperate regions. Stems (2-3 in. long), branches, and wavy leaves appear flattened. Stalk and capsule, often hidden by leaves, are both less than 0.2 in. long.

DELICATE FERN MOSS forms mats on earth and stones in moist places. Stems repeatedly branched and fernlike. Resembles Mountain Fern Moss (p. 110), but smaller overall. Few capsules are produced. Long-haired leaves surround reproductive organs. Ht. flat to 4 in.; L. 1-1.5 in.

WATER MOSSES grow submerged in clear brooks. Short-stalked capsules abundant on older, leafless parts of plant. Common Water Moss (stem to 40 in.; L. to 0.2 in.) is larger than similar, more widespread Fire-preventer Moss.

LIVERWORTS AND HORNWORTS

These plants are closely related to mosses and have a similar life cycle. Liverworts (8,500 species) may be leafy, like mosses, but the leaves are in two rows and flattened, although there may be a third row on the underside. The leaves are never in spirals. Leafy forms commonly occur on tree trunks or on rocks, usually with mosses and lichens. Other liverworts lack leaves and have a true thallus—flat, thin, ribbon-like, and often divided. A few of these are truly aquatic, but most grow on soil, wet rocks, or at the base of trees in swamps.

Hornworts (about 300 species) are often classified with the liverworts but are more advanced. The gametophyte is a flat, non-leafy thallus. The sporophyte develops as a spikelike capsule. The base of the capsule continues to grow throughout its life and in the center is a core of sterile cells that conduct food and water. Only three genera of hornworts occur in the U.S.

COMMON HORNWORT, found throughout the world, has a dark-green, lobed gametophyte, generally rough and without a midrib. Each cell of the thallus contains a single large chloroplast. Embedded in the thallus are the female reproductive organs—archegonia. Rodlike sporophytes that rise from the fertilized archegonia split in two, releasing the yellow spores and exposing a central columella. Fairly common on moist soils, in which it is anchored by rhizoids. Gametophyte 0.5-1 in. wide; sporophyte 0.5-1.0 in. high.

COMMON LIVERWORT, one of the largest, has a much-branched, ribbon-like thallus with internal air chambers forming a diamond-shaped pattern on outer surface. Male and female plants are separate. Each produces umbrella-like sex structures—the female nine-ribbed, the male smaller and disk-shaped. The pendant sporophytes develop later inside the archegonia under the female umbrella. Vegetative reproduction involves flat asexual buds (gemmae) on both female and male plants. Thallus about 0.5 in. wide; 2-5 in. long.

GREAT SCENTED LIVERWORT is similar to the Common Liverwort, but a pore is easily visible in the center of each hexagonal plate on the thallus. Sporophytes grow under the conical "umbrellas." This widespread species forms carpets on moist ground. Thallus to 0.5 in. wide; 2-5 in. long.

ASTERELLA is an aromatic liverwort often found on damp soil. Margins of the thallus and the scales underneath are purple. Sporophytes form under a lobed cap, each surrounded by a white fringe of scales. Thallus is about 0.3 in. wide; to 1 in. long.

COMMON SCALE MOSS (a liverwort) forms mats on wet soil. The wavy-margined thallus is thin, almost transparent when dry, with a conspicuous midrib along which sex organs form. Thallus is 0.2 in. wide and up to 4 in. long.

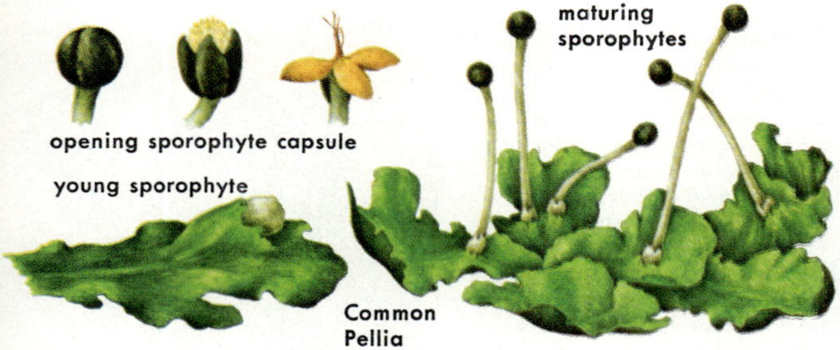

COMMON PELLIA (above) is a very common northern liverwort. Mature capsules shed spores, leaving a central tuft of hairs. Thallus, often branched and notched at the end, is up to 0.5 in. wide; 1 in. long.

LEUCOLEJEUNEA, also leafy, grows on trees and rocks. Upper leaves overlap. Part of each folds under, forming a series of toothed lobes that flank the round, central underleaves. Leafy stem 0.1 in. wide.

COMMON SCAPANIA appears to have four rows of leaves instead of two, as each leaf is bilobed. Margins strongly toothed. Brown gemmae. Grows in temperate N.A. and Europe. Leafy stem 0.5 in. wide.

SLENDER RICCIA is a floating thallus liverwort, the narrow, forked ribbons forming tangled masses. Sporophytes develop only when plants are stranded. Found world-wide. Thallus 0.04 in. wide; 0.5-2 in. long.

COMMON PORELLA has round, overlapping leaves. A lobe of each upper leaf folds under stem, which also has a series of oval underleaves. Forms mats on ground and tree trunks. Stem 0.1-0.2 in. wide; to 3 in. long.

COMMON FRULLANIA (one of more than two dozen N.A. species) forms dark designs on bark. Tiny leaves are closely overlapping and have saclike lobes underneath. Dry leaves barely visible. Leafy stem 0.1 in. wide.

THREE-LOBED BAZZANIA is one of the largest leafy liverworts. Leaf tips are three-toothed; underleaf coarsely toothed. Branches often upright. Forms mats in cool places; N.A. and Europe. Stem 0.5 in. wide.

PURPLE-FRINGED RICCIA has slim purplish scales under deeply furrowed, floating thallus. Less commonly grows on wet soils; is then ribbon-like and scaleless. Black sporophytes embedded in thallus. About 0.5 in.

LEAFY LIVERWORTS

Common Porella

lower surface

enlarged branch

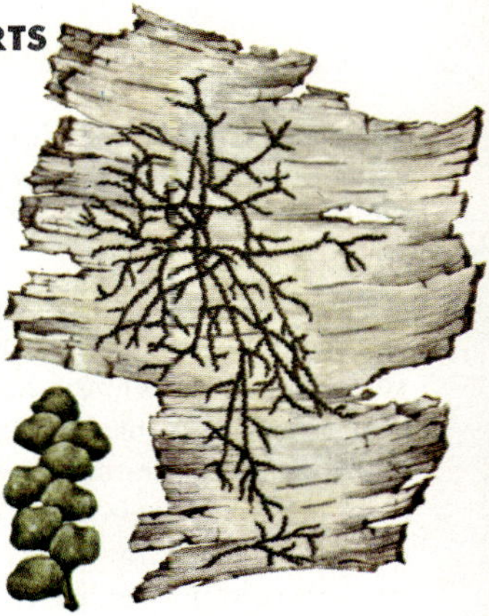

Common Frullania on birch bark

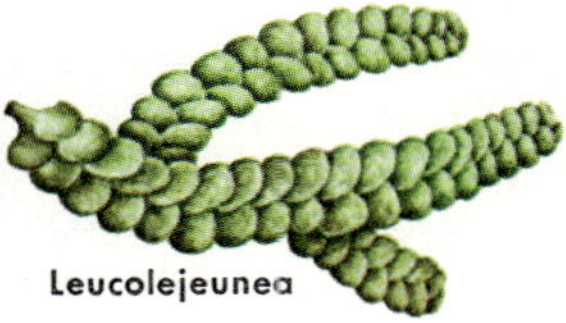

Leucolejeunea

upper lower

Common Scapania

Three-lobed Bazzania

AQUATIC THALLUS-TYPE LIVERWORTS

Slender Riccia

scale

Purple-fringed Riccia

FERNS AND FERN ALLIES

Ferns and their allies (about 10,000 species) are vascular plants that reproduce by spores rather than seeds. Ferns (p. 127-146) usually have large flat leaves on a stalk; fern allies (p. 118-126) have small, scalelike leaves and may resemble mosses, grasses or rushes.

Fern leaves or fronds unroll from curled fiddleheads. The leaves (below) may be either simple or compound—that is, divided into leaflets. A compound leaf may be divided once (pinnate), twice (bipinnate), or three times (tripinnate). Margins or edges may be entire, toothed, or lobed. When the clefts are deep and the lobes long and narrow, the margin is pinnatifid. In some ferns, a leaf may have a combination of characteristics, such as being pinnately compound at the base but pinnatifid near the apex.

Spores are usually produced on the underside of leaves in tiny capsules (sporangia) that are grouped in

FERN PARTS AND LEAF TYPES

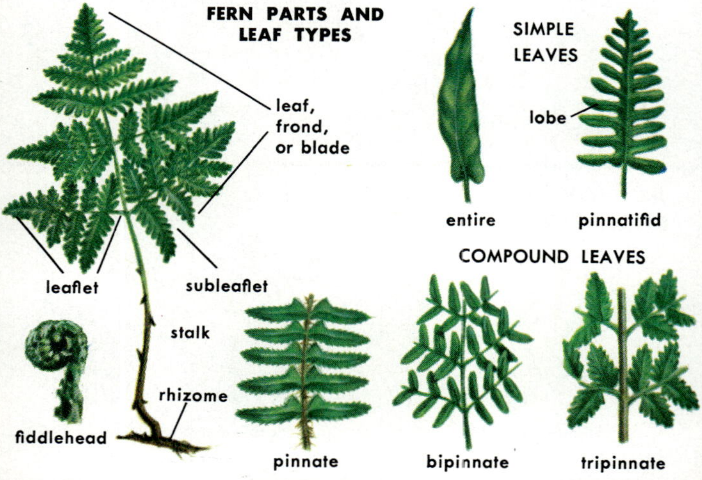

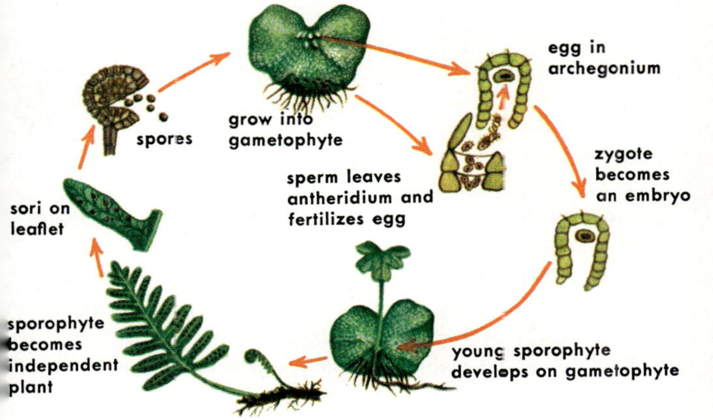

LIFE CYCLE OF A FERN

clusters (sori). These clusters may be under a cover (indusium). In some ferns, sporangia form on a separate fertile frond. Most ferns produce only one type of spore, but water ferns (p. 146) have two.

If a fern spore lands on soil or another suitable moist place, it germinates. The spore sends out a short tube that divides and grows into a small, green, flat, usually heart-shaped prothallus. This little plant, which lacks true roots, stems, and leaves, and resembles some liverworts, is the gametophyte, the fern's sexual stage. On its underside, sex organs develop: the female (archegonium) with an egg, and the male (antheridium) with sperms. The sperm swims in a film of water to fertilize the egg. The fertilized egg, or zygote, starts to grow, forming an embryo. Eventually small leaves appear, roots form, and the new plant soon becomes independent. This is the sporophyte, the fern we commonly see. Fern allies have similar life cycles.

FERN ALLIES consist of five families: whisk broom ferns, horsetails, quillworts, club mosses, and spike mosses. While their commercial value is negligible, these plants provide excellent ground cover, sheltering smaller plants and holding moisture in the soil.

Fern allies represent ancient groups of plants once much more numerous. Many are only slightly changed from but are often much smaller than their ancestors, which flourished 200 to 300 million years ago. Most coal deposits are the fossilized remains of giant ferns, towering horsetails, and many smaller ferns and fern allies. Fossils show the structure, history, and diversity of these early plants. They reveal, too, the existence, long ago, of many species now extinct.

Whisk broom ferns *(Psilotum)* are represented today by only two species, widely distributed in the tropics and subtropics, and by a relative growing in Australia and on nearby islands. The 25 living species of horsetails all belong to the genus *Equisetum*. Club mosses, spike mosses, and quillworts today number only about 1,000 species.

Fossilized leaves of a horsetail from coal beds

WHISK BROOM FERNS *(Psilotum)* lack roots. The stem is partially underground, and the tiny scalelike leaves alternate.

Found on humus and logs through tropical and subtropical regions, ranging north to South Carolina. Stem is twisted, wiry, and repeatedly forked. Three-lobed sporangia, yellow when ripe, are borne on the upper parts of branches. Ht. 6-18 in.

HORSETAILS *(Equisetum)* have ridged, segmented stems. The small, triangular leaves are united into a collar at each node. Spores form in a conelike strobilus that sometimes grows on a separate fertile stalk. Horsetails are world-wide in distribution.

DWARF HORSETAIL, the smallest species, grows in cool forests, forming curled, tangled mats. Stems solid, rarely branched. Silica deposits on their ridges give them a rough texture. The pointed, spore-bearing cones and the bases of the leaves are black. Common throughout northern N.A., but rare in New England. Ht. 6-10 in.

GIANT HORSETAIL is found near streams and swamps from southern California to British Columbia. Clusters of ivory-white to brown fertile stems from 1 to 2 ft. tall appear in early spring but soon shed their spores and wither. They are followed by taller (4-6 ft.), pale-green, hollow, sterile stems, with many lateral branches.

ROUGH HORSETAIL has so much silica in its hollow stems that it is often called Scouring Rush (as are some other species). The unbranched, rough, upright stems, with small, sharp-tipped cones and sheaths of toothlike leaves, grow to 5 ft. tall. Found throughout N.A., Europe, and Asia, in wet gravel or sand along streams, in moist woods, and on spoil banks.

SMOOTH HORSETAIL, a variable, smooth-stemmed species, is restricted to America and is most common in the Midwest and West. Grows in wet, sandy, and clay soils, especially along stream borders. Hollow stem rarely branches. Cones may be rounded or have sharp tip, as in Rough Horsetail. Sheaths of blackish teeth soon wither and fall. Ht. 2-4 ft.

FIELD HORSETAIL is widespread—the most common species in the Northern Hemisphere. It grows in woods and meadows and along roads and railroad tracks, preferring a moist habitat. The hollow sterile stems are green, with whorls of branches. Pink fertile stems with cones appear in early spring. This and other species are poisonous to livestock. Ht. 8-20 in.

WOOD HORSETAIL sends up fertile stems in spring. After spores are shed, whorls of green branches develop, like those on sterile stalks. This hollow-stemmed, woodland plant sends creeping rootstock deep into moist soil. One of the most beautiful horsetails, it occurs from Virginia to Nebraska northward to Montana; in northern Europe and Asia. Ht. 12-24 in.

MARSH HORSETAIL, rather rare in central N.A., is very common in northern Canada, Alaska, and colder parts of the Old World. It is found in marshy places, along streams and ponds, and in wet meadows. The hollow stems, usually less than 2 ft. tall, have upward-thrusting branches. The tip may be fertile or sterile. The strobilus is blunt, and about 1 in. long.

WATER HORSETAIL, common in sluggish waters of ponds, swamps, and ditches, grows around the world in a northern belt. Whorled branches that vary in length and number (sometimes completely absent) grow from sheaths after the cone is formed. Hollow stems have flat ridges and very little silica, making it one of the smoothest horsetails. Ht. 3-4 ft.

Water Horsetail

enlarged megaspore

BLACK-BASED QUILLWORT, dark at the base of the leaves, ranges from N.J. westward. Ht. 6-10 in.

BRAUN'S QUILLWORT, distinguished by its spiny megaspores, grows in northern N.A. and Europe. Ht. 5-10 in.

enlarged megaspore

QUILLWORTS *(Isoetes)* resemble young onions. The stiff, slender leaves, spoon-shaped at the base where they surround the bulblike stem, grow 6 to 20 in. tall. Most of the 65 species (19 in U.S.) grow partially or totally submerged; a few grow on land. Species are identified by sculpture of the megaspores (large female spores), produced in cases at the base of the leaves.

CLUB MOSSES *(Lycopodium)* are herbaceous plants with small leaves having a single unbranched stem. Spores are all of one size. Most are creeping evergreens. Over 100 species; nearly two dozen in N.A.

SHINING CLUB MOSS has erect stems about 6 in. high. The sharp-tipped leaves are often of unequal length and in ragged tufts. Spore-bearing leaves do not form a cone. Spores, as in all club mosses, are of uniform size. Grows in rich, acid soil of moist woods from eastern Canada west to Minn.; south to Ala.

FOXTAIL CLUB MOSS is semi-evergreen and bushy like a fox's tail. It grows up to 10 in. high from arched, creeping stems that turn yellow in winter. Spores form at base of leaves in the bushy, conelike tip. Found from Massachusetts to Uruguay, in acid bogs and swampy sand barrens, particularly near the coast.

STIFF CLUB MOSS ranges from Virginia into Canada; found locally in Mid-west and frequently on West Coast. A narrow yellowish cone tops most branches, which stick up stiffly (6 in.) from a wiry horizontal stem. Growth renews annually at tip, leaving a constriction between evergreen leaves of each branch.

RUNNING GROUND PINE is used for Christmas decorations. The rounded, branching pattern of the upright stems and the conspicuous clusters of long yellowish cones, 1 to 4 on each stem, make it easy to identify. Found in temperate regions of both hemispheres but more common on sandy and acid soils under conifers in North. Ht. 10 in.

WOLF'S CLAW CLUB MOSS, also used for Christmas decorations, is widely distributed and common in northern N.A., Europe, and Asia. Upright stems branched and densely set with leaves. Several 2-3-in. yellowish cones are found on nearly naked stems. Leaves hair-tipped; fertile leaf yellowish. Ht. to 1 ft.

TREE CLUB MOSS, or Ground Pine, has upright stems 8-12 in. tall, much-branched and treelike, with small, shiny leaves. Cones, yellow when mature, are 1-2 in. long. Usually found in open pine woods and bogs in central and northern N.A.

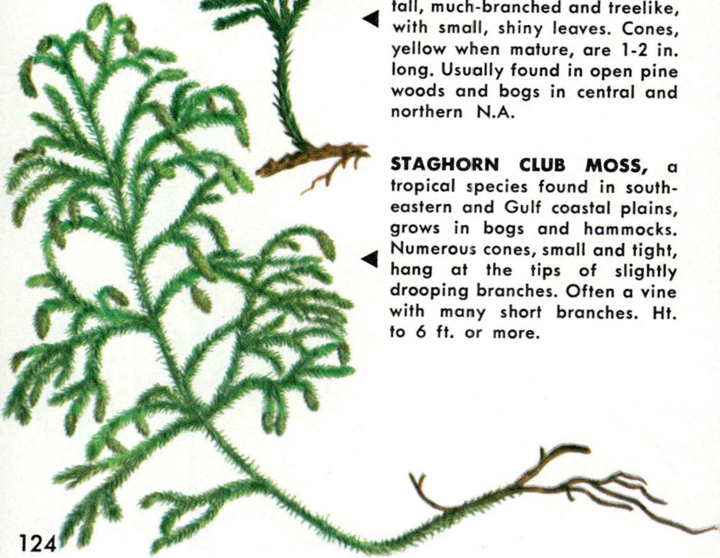

STAGHORN CLUB MOSS, a tropical species found in southeastern and Gulf coastal plains, grows in bogs and hammocks. Numerous cones, small and tight, hang at the tips of slightly drooping branches. Often a vine with many short branches. Ht. to 6 ft. or more.

SPIKE MOSSES *(Selaginella)* resemble club mosses but have two kinds of spores, large (megaspores) and small (microspores), borne in separate sporangia at the base of the spikelike leaves. About 700 species, mainly small or creeping plants, are widely distributed, especially in the tropics.

ROCK SPIKE MOSS is a small, gray-green, mosslike plant often growing in dense mats. Found in acid soils and on rocks in open areas. Crowded, overlapping narrow leaves form spirals along creeping stems. Short clusters of fertile leaves produced at ends of branches. Occurs South to Ala. and Cal. Ht. 3 in.

UNDERWOOD'S SPIKE MOSS, found on rock ledges of southwestern mountains, is more abundant in Colorado canyons. A sparsely branched, creeping plant (6 in. long), with stems that are densely covered with narrow, almost flat leaves.

RESURRECTION SPIKE MOSS, found in the arid Southwest and southward to Peru, rolls up when dry and uncurls when wet, thus remaining dormant during long dry periods. It grows again with each rainy season. Leaves are blunt. Sold as Resurrection Plant. Ht. 3 in., dry.

dormant plant *(left)*; moist plant *(right)*

FERN ALLIES

MEADOW SPIKE MOSS has small cones. Leaves near tips of branches are small, lance-shaped, and in four rows. Lower leaves on sides of stem are larger. Widely distributed from southern Canada to Gulf, growing in moist, open places. Creeping, 2-6 in.

BIGELOW'S SPIKE MOSS is a California species of scrubland, chaparral, and rocky foothills. Branches, dark green when young, rise 2 to 8 in. from long, creeping rhizomes. Scalelike leaves have a white bristle tip. Small cones plentiful at ends of short branches.

OREGON SPIKE MOSS, found from Oregon and Washington to Alaska, is an epiphyte, forming dense mats or hanging 6 in. to 3 ft. from trees in humid forests. The slender plant branches frequently. Light-brown stems are visible between the loosely spaced, narrow, pointed leaves.

FERNS have uncoiling leaves with branched veins. Their dustlike spores, all of one size except in water ferns, form in sporangia that are, in most ferns, grouped in sori formed as outgrowths on leaflets. Sori may also form on a fertile spike (adder's tongue) or be enclosed in sacs (water ferns). Ferns are most common in the tropics but are also plentiful in temperate regions. Many are plants of wet areas and shaded forests, but some grow in meadows, on cliffs, or as epiphytes. Shape and size vary. Average plant height or length of leaf (frond) is given for each species.

COMMON ADDER'S TONGUE has a single, simple leaf with an undivided blade, above which rises a fertile spike bearing two rows of sporangia. Veins in blunt-tipped leaf are netted and regular. Several varieties grow in damp fields and woodlands of temperate regions. Ht. 4-12 in.

LIMESTONE ADDER'S TONGUE is a similar but shorter fern. The larger, sharp-tipped leaf has irregular veins, some of them heavy. Found on shaded clay or gravel banks, limestone outcrops, pastures, and woodlands, chiefly in the West but also in southeastern states. Ht. 4-6 in.

Adder's tongue and grape ferns (p. 128) belong to a primitive family not closely related to other ferns. Rather than a flat prothallus, the gametophyte is tuberous and grows underground in association with fungi.

GRAPE FERNS

Moonwort

Grape Fern

MOONWORT, related to adder's tongue (p. 127), is one of the grape ferns, all of which bear sporangia in grapelike clusters on branched fertile spikes. Grows on rocky ledges, in dry meadows, and on hillsides of northern regions. Not common. Leaflets half-moon or fan-shaped. Ht. 6 in.

GRAPE FERN occurs in many varieties in eastern U.S. and less commonly in Midwest and Southwest. Sterile leaves are triangular, 3 in. long, lacy, and much-divided. Fertile spike, 6-10 in. tall, bears clusters of yellowish sporangia in autumn. Grows in pastures and open woodlands.

RATTLESNAKE FERN is large, conspicuous, and widely distributed in Northern Hemisphere. The single triangular leaf is divided into leaflets and lacy subleaflets. Fertile spike, bearing clusters of sporangia, appears in late spring or early summer and persists for several months. Ht. to 24 in.

Rattlesnake Fern

Rattlesnake

Grape

COMPARISON

ROYAL FERN and others on this page represent a family with large leaves that rise from a heavy rootstock, often forming in a tangled mat on the surface. Royal Fern leaves may be 6 ft. tall. Leaflets at the tip bear sori. Grows in wetlands from tropical South America into Canada. Also in Europe, Asia, and Africa.

CINNAMON FERN is common in wet places of Northern Hemisphere. Fertile leaves appear in spring; sori green at first, turn cinnamon as spores ripen. Broad, bipinnate sterile leaves produced later. Silvery hairs on fiddleheads and leaflet bases turn cinnamon. Ht. 3-4 ft.

INTERRUPTED FERN has fertile leaflets "interrupting" sterile leaflets in center of leaf. Dark brown when ripe, they wither and fall, leaving vacant spots along leaf stem after midsummer. Fiddleheads woolly; leaflets not hairy at base. Grows in rocky, dry soils of eastern N.A. Ht. 3-4 ft.

Royal Cinnamon Interrupted

OF SUBLEAFLETS

CLIMBING FERNS are a family with distinctly unfern-like leaves. Those of the Hartford Fern, which grows along the U.S. East Coast, are 3 ft. long, but each consists of numerous widely spaced, paired, palmate sterile leaflets with smaller, deeply lobed fertile leaflets topping vine. Another type is on p. 131.

FILMY FERNS are a family of seven species, all with thin, almost transparent leaves; fertile and sterile leaves alike. The sporangia, on a bristle-tipped stalk, have a cuplike indusium at the base. Each develops at the end of a vein. The rare Kraus' Filmy Fern (left), 1-2 in. long, grows in mats on rocks, logs, and trunks of trees in the Everglades. Bristle Fern (right), 6-8 in. tall, forms mats on damp rocks and ledges in the Appalachians. Overlapping leaves commonly form a dense mat.

sporangia

Related to the Hartford Fern is the Curly Grass Fern, locally abundant in New Jersey; rare elsewhere along the northeastern coast. Curly Grass Fern grows in grasslike tufts. It has coiled, twisted, threadlike sterile leaves, 2-3 in. long. Fertile leaves are longer, tipped with a divided blade.

POLYPODY FERNS, the largest fern family, includes the two ferns below and all those from p. 132 to p. 145. The genus *Polypodium* is widespread. Coast Polypody, a western species, has leathery pinnatifid leaves, 5-20 in. long. The leaf lobes are broad and blunt, with round sori along the midrib. It grows on tree trunks, logs, and ledges. Licorice Fern, with similar range and growth habit, has thin, tapered leaves with pointed lobes. Its rootstock has a licorice flavor.

Coast
Polypody

Licorice
Fern

sori on
lobe

STRAP FERN, one of many tropical species that extend into Florida, grows abundantly in hammocks on logs, stumps, and tree trunks. Straplike leaves often 2 to 3 in. wide and up to 4 ft. long; prominently veined. Veins form diagonal rows from midrib; sori along veins on underside.

RESURRECTION FERN, a common polypody, has leaves that curl and turn brown when dry, revive when moist. Grows on trees and rocks from tropics through eastern N.A. Ferns at bottom of p. 131 are also in genus *Polypodium*. Ht. 1-7 in.

COMMON POLYPODY grows on logs or rocks in damp, shady places in cool temperate areas. Large, naked, red-brown sori prominent on underside of leaflets; 10 to 20 pairs. Creeping rhizome is covered with reddish scales. Ht. 6-12 in.

LEATHER FERN, a world-wide tropical species, is prominent near mangroves and in marshes throughout southern Florida. The stiff, erect, compound leaves may be over 9 ft. long. Masses of sori cover the undersurface of many leaflets near the tip of the leaf with a reddish-brown, felt-like mat.

GOLDEN SERPENT FERN grows on palms or other trees and on ground. Its creeping, serpent-like rhizome is fuzzy, with red scales. Deeply cut leaves have golden sori on the underside. Common from tropics to northern Florida. Ht. 10-30 in.

VINE FERN is a delicate tropical American fern found in Florida. Tapered leaves are loosely spaced along a rhizome that clings to trunks or branches by many small roots. Sori are aligned on each side of midrib. Leaf 2-6 in. long.

SHOESTRING FERN has long (4-40 in.) grasslike leaves with sori underneath in two marginal grooves, sometimes covered by leaf margins. Grows in drooping clusters from scaly rhizome. Common in West Indies and tropical America and on palm trunks in southern Florida.

LADDER BRAKE is an Old World fern now growing widely as an escape in warm parts of the Americas. Leaf bipinnate, with sori lining the undermargin of the narrow leaflets and covered by rolled-under edge. Veins fork near midrib and sometimes again near margin. Ht. 12-18 in.

COMMON BRACKEN forms dense growths on poor soils of temperate woods and fields. A world-wide species. Leaves are coarse, strongly triangular, with sori lining margins of subleaflets. Thick rootstock spreads 10 ft. or more each year. Ht. to 6 ft.

PURPLE CLIFF BRAKE is found on dry limestone cliffs, rocks, and masonry throughout N.A. Leaves have brittle purple stalks and widely spaced leaflets that may, on the same leaf, be further divided into subleaflets. Rolled-under leaflet margin covers sori. Ht. 12 in.

VENUS MAIDENHAIR FERN encircles the world in the tropics and warm temperate regions. It grows on moist limestone rocks and rock ledges. Leafstalk, shiny black, lacks the curved branching found in the Maidenhair Fern (below). The subleaflets are fan-shaped, with modified tips that cover moon-shaped sori. L. 10-18 in.

MAIDENHAIR FERN, one of the most beautiful ferns, has finely divided leaves 4-16 in. wide, almost horseshoe-shaped. Subleaflets vary from fan-shaped to oblong; sori are covered by the tips. The 10-20-in., wiry leafstalk is shiny black. Known throughout eastern U.S., in Canada, and in the Northwest.

fertile leaflets

EBONY SPLEENWORT has long, stiff, ladder-like fertile leaves, with shiny brown stalks and narrow leaflets, "eared" at the base. Sori in pairs along midrib. Sterile leaves shorter and less erect. In rocky soils, woods, fields, and along walls in eastern N.A. west to the Rockies and south to Gulf. Ht. 7-18 in.

MAIDENHAIR SPLEENWORT has fertile leaves that grow erect; sterile leaves often flat. Leaflets are rounded, with a few blunt teeth; leafstalks brittle and wiry. Found on rocks and ledges, often in the open. A widespread fern in central and eastern N.A. Ht. 6-8 in.

fertile leaflet

WALL RUE is a small fern, growing on limestone rocks or walls throughout eastern N.A. The Old World has a very similar species. Leaves bipinnate or tripinnate; delicate but not fernlike. Leaflets and subleaflets have a long stem. Sori are often united along the veins of fan-shaped subleaflets. Ht. 3-5 in.

fertile subleaflet

fertile lobes

SILVERY SPLEENWORT has light-green, hairy leaves that grow 3 ft. tall or more in rich, moist woods. Leaflets deeply lobed, with toothed or smooth margins. Narrow sori are covered with indusia that are silvery but later turn reddish brown. Found in central and eastern N.A. and in Asia.

LADY FERN is large but strikingly delicate, with variable leaflets. Sori are covered by kidney-shaped, hairy indusia. This lowland fern is found in moist places through much of N.A. and temperate parts of the Old World. Ht. 12-36 in.

HALBERD FERN is a tropical American fern of south and central Florida and Texas, found especially in limestone sinkholes. Stout brownish rhizome. Leaves triangular or halberd-shaped, usually lobed with three to five leaflets. Numerous round indusia are scattered on the back of leaflets. Ht. 12-20 in.

indusia on underside of leaflet

CRESTED FERN is a blue-green shield fern (indusium shield-shaped) with widely spaced leaflets that commonly twist to assume a horizontal position. Sori are on upper leaflets only, midway between margin and midvein. It is found in swampy woods of central North America, Europe, and Asia. Ht. 2-4 ft.

MALE FERN, common in wooded areas of Europe, is also found, though less abundantly, in northeastern N.A. and in Asia. Leaf is broadest above middle, narrowed at base. Sori, near midvein of toothed subleaflets or lobes, are covered by kidney-shaped indusia, commonly hairy around the edges. Ht. 1.5-2 ft.

MARGINAL WOODFERN is similar to the Male Fern, but the sori are marginal and the fronds a darker, bluer green. Subleaflets only slightly toothed. Fiddleheads are densely covered with golden-brown, furry scales. Found on rocky woodland slopes of central and eastern N.A. Ht. 1.5-2 ft.

AMERICAN TREE FERN is a shield fern, not a true tree fern. Its erect trunk, a foot or two high, supports a crown of arching leaves about 6 feet long. Limited to the West Indies, tropical America, and Florida, it is widely cultivated as an ornamental. Sori are exposed when short-lived indusia disappear.

MARSH FERN grows in marshes, wet woodlands, and damp meadows but seldom in standing water. Lower leaflets are opposite; those above may be alternate. Fertile leaflets, mostly toward leaf tip, have margins curved over sori. Leafstalk dark at base. Eastern and central N.A., Europe, and Asia. Ht. 1.5-2.5 ft.

NEW YORK FERN, a tapering fern resembling the Marsh Fern, grows in somewhat drier areas. The lower two to four pairs of leaflets are greatly reduced in size, and leaflets are never opposite as in Marsh Fern. Found in eastern and central N.A. A related species grows in California. Ht. 1-2 ft.

LONG BEECH FERN is found in northern N.A., south in mountains to Virginia and Oregon; also in Europe and Asia. This fern grows on wet rocks, cliffs, and shaded banks. Drooping basal leaflets not winged; widely spaced from those above. Upper leaflets joined to leaf stem with a broad, winged base. Sori on lower leaflets; no indusium. Ht. 1-1.5 ft.

HOLLY FERN is found in northern N.A. to Michigan, and in western mountains; also in northern Europe. Grows chiefly on limestone ledges and rocky slopes. Evergreen leaves taper in both directions. Eared leaflets with bristle-tipped teeth tend to be sickle-shaped. Cinnamon scales cover base of leaf stalk. Sori large, in two rows. Ht. 1-2 ft.

CHRISTMAS FERN is an evergreen sometimes used in Christmas decorations. Stiff leaves are once divided, with spiny and distinctly eared leaflets. Larger leaves have fertile tips with numerous round sori in two or more rows near the midvein. Leafstalk is scaly. Christmas Fern grows through much of eastern N.A. Ht. 1.5-2.5 ft.

BLADDER FERN is chiefly a northern plant in N.A. and Europe, extending southward only in mountains. Found in rock crevices, in soil, and on tree stumps in moist woods. Brown hairs or scales may cover rootstock. Leaves widest below the middle, tapering gradually to the tip. Sori are few and scattered, at first covered by bladder-like indusia. Ht. 1-2 ft.

BULBLET FERN leaflets are variable in shape, drooping at the ends. Tiny bulblets that appear on lower surface of fertile subleaflets fall off when mature and produce new ferns. Reproduces also by spores. Sori are under hoodlike indusia on subleaflet veins. Widespread in central and eastern N.A. Fertile leaves to 5 ft. long.

HAY-SCENTED FERN, or Boulder Fern, grows so abundantly in dry woodlands and pastures that it is often listed as a weed. It has a lacy appearance and haylike fragrance. The sharp-tipped, hairy leaflets are close together and not all opposite. Small sori are along indentations of subleaflets. Found in eastern and Central U.S. Ht. 1-3 ft.

▲ **BOSTON FERN** occurs wild in southern Florida, the West Indies, and tropical America, growing in humus, in limestone sinkholes, or as an epiphyte. The leaves may be 3 ft. or longer, with eared leaflets. Sori are covered with a kidney-shaped indusium. Many varieties are cultivated as house plants.

VIRGINIA CHAINFERN grows in bogs and lowland woods of central and eastern N.A. Thick, erect leaves with dark stalks have tapered narrow leaflets that are deeply cleft. On the underside are numerous oblong indusia-covered sori, in double rows along veins of both leaflets and lobes. Ht. 3-5 ft.

NETVEIN CHAINFERN has raised veins forming a conspicuous net. Pinnatifid sterile leaf resembles Sensitive Fern's (p. 145) but with lobes slightly offset rather than opposite, and margins not as wavy. Sori on narrow fertile leaves are oblong with leathery indusium. Central and eastern U.S. Ht. 2-6 ft.

▲

HART'S-TONGUE FERN is rare and local in eastern N.A. Some are nursery escapes. American form is similar to the more common European variety but not as hardy. Leaves are dark and thick, with wavy edges. Sori are long and in pairs alongside veins. Grows on moist limestone. Ht. 6-18 in.

OSTRICH FERN has 5-ft. tufts of dark-green sterile leaves, widest near the top and tapering at the base. Fertile leaves are shorter and stiff, with leaflets that roll into the midrib and enclose the sori in a "pod." Grows in swamps and moist, open woods from Virginia to the Arctic Circle; also in Europe and Asia.

OAK FERN, 5 to 10 in. tall, resembles a tiny bracken (p. 134). Each frond has three triangular leaflets—the lower two opposite, as are the subleaflets. Small, round, naked sori develop near the margin of subleaflet lobes. Grows in cool, moist woods from Virginia northward; also in western mountains, Europe, and Asia.

BLUNT-LOBED WOODSIA belongs to a large and varied group of ferns. This small species is found on shaded limestone ledges. Leaflets are spaced widely; subleaflets bluntly lobed. Stems are hairy. Sori round and held in the indusium, which splits open like flower petals. Grows in clusters from stout rootstocks. Widespread in N.A. from Alaska and Maine south to Georgia. Ht. 12-15 in.

OREGON WOODSIA is mainly a western species but is found scattered in moist, rocky places of northern and central N.A. Leaf blades are narrow and hairless. Leaflets short and blunt. Stalk brown. The indusium breaks into hairlike lobes when it releases the sori. One of 37 species of Woodsias of cooler areas of the world. Ht. 14 in.

WOOLLY LIPFERN curls and shrivels when dry but becomes a furry, tufted fern when moist. Brown stalk hairy and scaly. Densely woolly, tripinnate leaves become reddish, especially underneath. Round marginal sori covered by rolled leaf edge. On exposed ledges from Mexico and Arizona to Missouri and Virginia. Ht. 6-8 in.

SLENDER LIPFERN, a graceful western relative of Woolly Lipfern (p. 144), grows from British Columbia to Mexico and occurs locally in the Midwest. The leafstalks are long, slim, and hairy. Slender, tapered leaves are gray-green, with whitish hairs above and densely covered with coarse brown hairs below. Leaflets distinctly rounded, with sori covered by inrolled margins. Ht. 8-12 in.

WALKING FERN has long, narrow, arching leaves that sprout new plants where their tips touch the ground, thus "walking" and spreading. Sori on undersurface of leaf are scattered along the network of veins. Grows on moist, shaded limestone rock or soil, but nowhere common. From Quebec to Minn. to Ga., Okla. Ht. 8-14 in.

SENSITIVE FERN has long, coarse, triangular sterile fronds, pinnatifid or divided into leaflets at base of the leaf. The leaflets on short (1 ft.) fertile fronds are small, hardened, and beadlike, enclosing the sori. The Sensitive Fern is found in damp places from Canada to the Gulf. It wilts rapidly when picked. Ht. 1.5-2 ft.

WATER FERNS

FLOATING MOSS FERN, about 1 in. across, has paired, rounded, hairy leaves that float and a third that is threadlike, bearing sporocarps. Stem between individual plants breaks easily, each part continuing independent growth. Probably native to Eurasia but found in tropics and scattered in N.A.

MOSQUITO FERN, one of the water ferns, is a small, free-floating plant with leaves about 0.3 in. long, each made of two oval leaflets. Plants form thick, greenish or reddish sheets on ponds. Spore cases, in hard sporocarps, contain two types of spores—small microspores (male) and large megaspores (female). Found along Atlantic and Gulf coasts and in the Mississippi Valley.

WATER SHAMROCK roots in the mud of pond bottoms. Note the clover-like, four-parted, floating leaves. Sporangia are in hard sporocarps. Introduced from Europe, it has spread through northeastern U.S. Closely related species are native in South and West. Leaf is 0.5-1 in. across; stalk 6-10 in. long.

Naked seeds of gymnosperms form on an exposed surface, as on top of pine-cone scales. Seeds of angiosperms (flowering plants) are surrounded by ovary wall that develops into a fruit, as in cherry.

GYMNOSPERMS

All of the non-flowering seed plants are gymnosperms. The name means "naked seed" and refers to the fact that the seeds are not enclosed in fruits, as they are in the flowering plants. The spore-bearing leaves are produced in cones of two kinds—male and female. In some, such as junipers, podocarps, and yews, the cones may be fleshy and berry-like but nevertheless they are not true fruits.

All of the gymnosperms are woody plants. Most become trees, although some are shrubs and a very few are climbers. The leaves may be large and fernlike, as in the cycads, or may be needles, scales, or narrow and flattened, as in conifers. Most gymnosperms are evergreen. In North America, only Baldcypress, larches, and a few others are deciduous, losing all of their leaves before growing a new set.

Living gymnosperms are classified in four orders: Cycads (p. 149), containing fewer than a hundred species, which grow in the tropics and subtropics; the Ginkgoes (p. 150), with a single species native to China—now widespread but known only in cultivation; the Ephedras and their kin (p. 150), a group of unusual and distinctly different plants, sometimes considered transitional to the angiosperms; and Conifers (p. 151), the largest and most important order of gymnosperms, with more than 500 species, found mainly in cool or mountainous regions. Conifers supply most of the world's lumber and pulp, as well as resins from which turpentine and aromatic oils are obtained.

Cycads and the ginkgo produce motile male cells; ephedras and conifers, wind-distributed pollen.

LIFE CYCLE OF PINE

In spring, ovule in female cone forms gametophyte containing egg. Male cone produces pollen with sperm nucleus. At pollination, sperm nucleus fertilizes egg. Fertilized egg (zygote) becomes an embryo within the seed, which forms on cone scale. Seed falls to ground when mature. Some seeds germinate and grow into new cone-bearing plants.

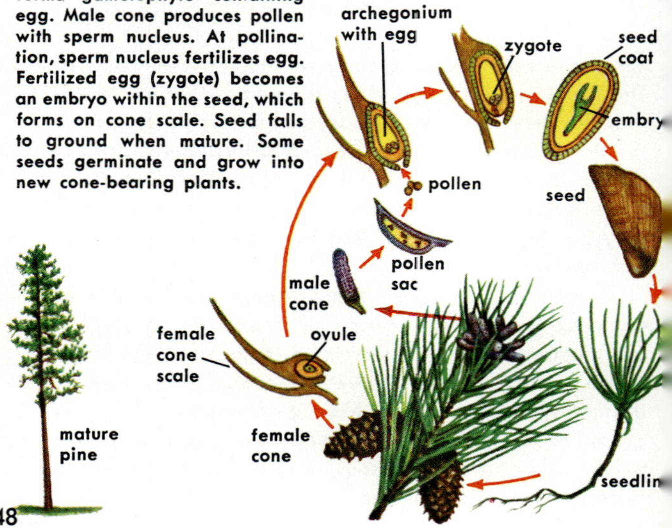

CYCADS are an old group of gymnosperms, mostly extinct. Nine genera grow in the tropics and subtropics. All have large, fernlike leaves that form a crown at the top of the usually unbranched stem. Male and female cones are borne at the crown, but on separate plants.

Coontie, or Zamia, is a small but common cycad of Florida pinelands. The stem is mostly underground, and the stubby end that emerges bears a crown of leathery, compound leaves. Indians and early settlers made flour from the underground stems. Coontie is poisonous when fresh, but the water-soluble poison was leached or washed out as the stems were ground.

Sago Palm, not a palm but a cycad, is a native of the Orient but is commonly cultivated in greenhouses and out-of-doors in Florida, California, and Hawaii. The large male cones give off an offensive odor, but they may be removed without harm to the trees.

GINKGO, or Maidenhair Tree, represents an order known for many years only as a fossil, from rocks as old as 200 million years. Then it was discovered that Tibetan monks were growing the trees in gardens of temples and shrines. The Ginkgo is now planted widely as a street and shade tree, 50-80 ft. tall. The female trees are often destroyed because their flesh-covered seeds have a rancid odor. The deciduous, bilobed leaves, borne on stubby stems, are wedge-shaped, resembling those of the Maidenhair Fern.

EPHEDRAS represent a small order of only three genera. Ephedras are shrubby (2-6 ft.) plants of the southwestern desert with jointed green stems (ephedra means horsetail). Small, scalelike leaves form at the stem nodes on new growth but soon dry up and are shed. Indians made a medicinal beverage, Mormon tea, from the plants. Green Ephedra, which grows in the mountains, has bright-green, awl-shaped, opposite leaves. Three-leaved Ephedra, usually found in rocky or sandy lowlands, has pale-green leaves mostly in whorls of three.

CONIFERS, mainly trees but a few shrubs, grow predominantly in cool temperate regions, though some are found in the tropics. Nearly all are evergreen, holding their needles or scalelike leaves for many years and never shedding all at once. Only Baldcypress and larches among North American conifers are deciduous. Most bear conspicuous cones, except for the podocarps, yews, and junipers which have small, fleshy cones. Common North American conifers—pines, spruces, firs, junipers, redwoods, Baldcypress, and others—are described in the Golden Nature Guide TREES. Five less common conifers and two exotics are described here.

ATLANTIC WHITE CEDAR, found from Maine to Florida and Mississippi, has flattened, nondrooping branchlets. Related species—Lawson's Cypress and Nootka Cypress—grow on West Coast. To 40 ft.

ARIZONA CYPRESS is a true cypress found in Arizona, New Mexico, and Mexico. This beautifully shaped tree grows to 40 feet tall and is planted as an ornamental in the Southwest.

INCENSE CEDAR, member of another widespread group, is native to Pacific Coast. Leaves scalelike; cones oblong. Grown as an ornamental. To 100 ft.

Florida Torreya

Eastern Yew

FLORIDA TORREYA, or Stinking Cedar, is a rare tree (to 40 ft. tall), found only in northwestern Florida. Its narrow, yewlike leaves have a fetid odor when bruised. California Torreya is the only other N.A. species. Two others grow in Asia.

MONKEY-PUZZLE TREE, to 50 ft. tall, has scalelike leaves on stout branches in whorls of five. A native of Peru, this araucaria is planted farther north than the Norfolk Island Pine and other araucarias that are grown as warm-area ornamentals.

EASTERN YEW, or Ground Hemlock, is a shrubby plant (4-6 ft.) that bears its seeds in a fleshy cup. Western Yew grows to 60 ft. tall and is commonly planted as an ornamental, as is the introduced and widely planted English Yew.

YEW PODOCARPUS and many other podocarps are grown as ornamental trees (to 60 ft.) or shrubs; sometimes as hedges, in California and southern U.S. Podocarps are native to the Southern Hemisphere, mainly in the tropics and subtropics.

Monkey-puzzle Tree

Yew Podocarpus

MORE INFORMATION

Botany textbooks are a good source of information about non-flowering plants. Some widely used ones are listed below, followed by other recommended reading. Accurate identification of non-flowering plants often involves details beyond the scope of this introductory volume, but more comprehensive books from the list below will help.

Botany Texts

Cronquist, Arthur, INTRODUCTORY BOTANY. Harper & Bros., New York, 1961.

Russell, Norman H., INTRODUCTION TO THE PLANT KINGDOM. C. V. Mosby Co., St. Louis, Mo., 1958.

Wilson and Loomis, BOTANY. Holt, Rinehart & Winston, New York, 1962.

Books on Specific Groups

Christensen, Clyde M., COMMON FLESHY FUNGI. Burgess, Minneapolis, Minn., 1959.

Cobb, Boughton, A FIELD GUIDE TO THE FERNS. Houghton Mifflin, Boston, 1956.

Conard, Henry S., HOW TO KNOW THE MOSSES AND LIVERWORTS. W. C. Brown, Dubuque, Iowa, 1956.

Fink, Bruce, LICHEN FLORA OF THE UNITED STATES. Univ. of Michigan Press, Ann Arbor, 1960.

Frye, Theodore C., FERNS OF THE NORTHWEST. Binfords & Mort, Portland, Oregon, 1956.

Grout, A. J., MOSSES WITH A HAND-LENS. 4th Ed., pub. by author, 1947.

Hale, Mason E., Jr., LICHEN HANDBOOK. Smithsonian Inst., Washington, D. C., 1961.

Prescott, Gerald W., HOW TO KNOW THE FRESH-WATER ALGAE. W. C. Brown, Dubuque, Iowa, 1954.

Ramsbottom, John, MUSHROOMS AND TOADSTOOLS. Collins, London, 1959.

Smith, Alexander H., THE MUSHROOM HUNTER'S FIELD GUIDE. Univ. of Mich. Press, Ann Arbor, Mich., 1963.

Stokoe, W. J., OBSERVER'S BOOK OF FERNS. Warne & Co., London.

Wherry, Edgar T., THE FERN GUIDE. Doubleday & Co., Garden City, N. Y., 1961; also THE SOUTHERN FERN GUIDE, 1964.

SCIENTIFIC NAMES

8 Red alga: *Chondrus* sp.
Yellow-green: *Botrydium* sp.
Mushroom: *Morchella esculenta*
Lichen: *Cladonia cristatella*
9 Moss: *Funaria* sp.
Liverwort: *Ricciocarpus* sp.
Fern: *Thelypteris palustris*
Club: *Lycopodium carolinianum*
Cycad: *Zamia floridana*
11 Sea Lettuce: *Ulva lactuca*
Gill: *Schizophyllum commune*
13 *Nectria cinnabarina*
16 *Caulerpa prolifera*
Merman s: *Penicillus dumetosus*
Sea Lettuce: *Ulva lactuca*
Mermaid's: *Acetabularia crenulata*
Sargassum vulgare
Rockweed: *Fucus vesiculosus*
17 *Ceramium fastigiatum*
Irish Moss: *Chondrus crispus*
Dasya pedicellata
Kelp: *Laminaria agardhii*
Padina sanctae-crucis
Polysiphonia nigrescens
18 Mermaid's Hair: *Lyngbya majuscula*
Chroococcus sp.
Oscillatoria limosa
Merismopedia elegans
Nostoc pruniforme
Gloeotrichia echinulata
19 *Batrachospermum moniliforme*
Synura uvella
Dinobryon divergens
20 *Vaucheria sessilis*
Tribonema bombycinum
Botrydium granulatum
Pinnularia borealis
Tabellaria flocculosa
Asterionella gracillima
Cyclotella meneghiniana
21 *Gymnodinium brevis*
Gonyaulax palustre
Ceratium hirundinella
Euglena viridis
Phacus acuminatus
22 *Volvox globator*
Draparnaldia glomerata
Ulothrix zonata
Oedogonium crispum
23 *Spirogyra communis*
Closterium moniliferum
Chlorella pyrenoidosa
Cosmarium sp.
Pediastrum boryanum
Staurastrum paradoxum
Chara sp.
24 Green: *Penicillium digitatum*
Corn: *Ustilago maydis*

24 (cont.)
Water: *Saprolegnia parasitica*
Black: *Guignardia bidwellii*
Bact.: *Pseudomonas solanacearum*
Shelf: *Ganoderma applanatum*
Cup: *Plectania coccinea*
25 Ringworm: *Microsporon lanosum*
Mushroom: *Clitocybe* sp.
27 Hay: *Bacillus subtilis*
Blue: *Pseudomonas syncyanea*
Tetanus: *Closteridium tetani*
Diphtheria: *Corynebacterium diphtheriae*
Pneumonia: *Diplococcus pneumoniae*
Scarlet: *Streptococcus pyogenes*
Sarcinia lutea
Spirillum rubrum
Syphilis: *Treponema pallidum*
28 *Physarum polycephalum*
Serpent: *Hemitrichia serpula*
29 Deep: *Stemonitis splendens*
Wood.: *Lycogala epidendrum*
Basket: *Dictydium cancellatum*
30 *Fuligo septica*
Crys.: *Didymium squamulosum*
Gray: *Physarum cinereum*
Green: *P. viride*
31 *Ceratiomyxa fruticulosa*
Ash-gray: *Arcyria cinerea*
Variable: *Trichia varia*
Common: *Badhamia macrocarpa*
33 Bread Mold: *Rhizopus stolonifer*
Water: *Saprolegnia parasitica*
Downy: *Phytophthora infestans*
White Rust: *Albugo candida*
34 Yeast: *Saccharomyces cervisiae*
35 *Penicillium notatum*
Aspergillus niger
Powdery: *Microsphaeria alni*
36 Common: *Morchella esculenta*
Brain: *Gyromitra esculenta*
Elastic: *Helvella elastica*
37 Summer: *Tuber aestivum*
Smooth: *Geoglossum difforme*
Orange Cup: *Aleuria aurantia*
Red Cup: *Plectania coccinea*
38 Fire: *Pyronema confluens*
Helotium citrinum
Monilinia fruticola
Coryne sarcoides
39 *Cordyceps militaris*
Daldinia concentrica
Dead Man's: *Xylaria polymorpha*
Ergot: *Claviceps purpurea*
42 Corn Smut: *Ustilago maydis*
43 Wheat Rust: *Puccinia graminis tritici*

44 Cedar Apple: *Gymnosporangium juniperi-virginianae*
Bean Rust: *Uromyces phaseoli*
White: *Cronartium ribicola*
Asparagus: *Puccinia asparagi*
45 Clammy: *Calocera viscosa*
Ear: *Auricularia auricula-judae*
Orange: *Tremella mesenterica*
46 Straight: *Clavaria stricta*
Ashy Coral: *C. cinerea*
47 Hedgehog: *Hydnum erinaceus*
Repand Hydnum: *H. repandum*
Horn: *Craterellus cornucopioides*
Hairy: *Stereum hirsutum*
48 Edible: *Boletus edulis*
Lurid: *B. luridus*
Yellow: *B. subtomentosus*
Pine: *Strobilomyces strobilaceous*
49 Sulphur: *Polyporus sulphureus*
Multi.: *Polystictus versicolor*
Red: *P. sanguineus*
Fir: *P. abietinus*
50 Artist's: *Ganoderma applanatum*
Rusty-hoof: *Fomes fomentarius*
Dry-rot: *Poria incrassata*
Oak: *Daedalea quercina*
52 Spring: *Amanita verna*
Death Cap: *A. phalloides*
Destroying Angel: *A. virosa*
53 Fly Amanita: *A. muscaria*
Panther: *A. pantherina*
54 Blusher: *A. rubescens*
Caesar's: *A. caesarea*
55 Sheathed: *Amanitopsis vaginata*
Volvate: *A. volvata*
Strangulated: *A. strangulata*
56 Honey: *Armillaria mellea*
Slimy: *A. mucida*
58 Morgan's: *Lepiota morgani*
Parasol: *L. procera*
Shaggy: *L. rhachodes*
Crested: *L. cristata*
59 Smooth: *L. naucinoides*
American: *L. americana*
Shield: *L. clypeolaria*
Yellow: *L. lutea*
60 Leopard: *Tricholoma pardinum*
Soapy: *T. saponaceum*
Red-haired: *T. rutilans*
61 Masked: *T. personatum*
Gray: *T. terreum*
Naked: *T. nudum*
Equestrian: *T. equestre*
62 Yellow: *Clitocybe aurantiaca*
Adirondack: *C. adirondackensis*
Sweet-scented: *C. odora*
63 Jack-o'-lantern: *C. illudens*
Clouded: *C. nebularis*
Sudorific *C. sudorifica*
64 Waxy: *Laccaria laccata*
Amethyst: *L. amethystina*
Purplish.: *L. ochropurpurea*
65 Velvet.: *Collybia velutipes*
Rooting: *C. radicata*
66 Broad-gilled: *C. platyphylla*
Buttery: *C. butyracea*
Tufted: *C. acervata*
Little Wheel: *Marasmius rotula*
67 Black-stemmed: *M. androsaceus*
Garlic: *M. scorodonius*
Acrid: *M. urens*
Fairy Ring: *M. oreades*
68 Capped: *Mycena galericulata*
Bleeding: *M. haematopus*
Clean: *M. pura*
69 Meadow: *Hygrophorus pratensis*
Conical: *H. conicus*
Vermilion: *H. miniatus*
70 Elm Tree: *Pleurotus ulmarius*
Oyster: *P. ostreatus*
Sapid: *P. sapidus*
71 Scaly: *Lentinus lepideus*
Astringent: *Panus stipticus*
Common: *Schizophyllum commune*
72 Deli.: *Lactarius deliciosus*
73 Sweetish: *L. subdulcis*
Orange-brown: *L. volemus*
Brown Velvet: *L. lignyotus*
Indigo: *L. indigo*
Woolly: *L. torminosus*
Common: *L. trivialis*
74 Emetic: *Russula emetica*
Fetid: *R. foetens*
75 Blackening: *R. nigricans*
Encrusted: *R. crustosa*
Yellow: *R. ochroleuca*
Green: *R. virescens*
76 Birch: *Lenzites betulina*
Brown: *L. saepiaria*
Chan.: *Cantharellus cibarius*
Vermilion: *C. cinnabarinus*
77 Fawn-colored: *Pluteus cervinus*
Silky: *Volvaria bombycina*
Handsome: *V. speciosa*
78 Livid: *Entoloma lividum*
Steel-blue: *Leptonia chalybaea*
Abortive: *Clitopilus abortivus*
79 Violet: *Cortinarius violaceus*
Earth-leaf: *Inocybe geophylla var. lilacina*
Com.: *Naucoria semiorbicularis*
80 Yellow: *Pholiota aurea*
Gypsy: *P. caperata*
Sharp-scale: *P. squarrosoides*
Early: *P. praecox*
81 Meadow: *Agaricus campestris*
Horse: *A. arvensis*
82 Bleeding: *A. haemorrhoidaria*
Sylvan: *A. silvicola*
Flat-capped: *A. placomyces*

83 Verdigris: *Stropharia aeruginosa*
Brick.: *Hypholoma sublateritium*
Tufted Yellow: *H. fasciculare*
84 Shaggy Mane: *Coprinus comatus*
Com. Ink-cap: *C. atramentarius*
Glistening Ink-cap: *C. micaceus*
85 Common: *Psathyrella disseminata*
Bell.: *Panaeolus campanulatus*
Butterfly: *P. papilionaceus*
86 Giant: *Calvatia maxima*
87 Gem.: *Lycoperdon gemmatum*
Pear-shaped: *L. pyriforme*
Beautiful: *L. pulcherrimum*
Spiny: *L. echinatum*
88 Many.: *Pisolithus tinctorius*
Stalked: *Tulostoma mammosum*
Triplex: *Geaster triplex*
Crowned: *G. coronatus*
89 Common: *Crucibulum vulgare*
Striate: *Cyathus striatus*
White: *Nidularia candida*
90 Small: *Laternea triscapa*
Lattice: *Clathrus cancellatus*
Columnar: *C. columnatus*
91 Dog: *Mutinus caninus*
Common: *Phallus impudicus*
Collared: *Dictyophora duplicata*
92 *Alternaria* sp.
Fusarium sp.
Helminthosporium sp.
Collectotrichum sp.
Pestalotia sp.
93 Blister Lichen: *Physcia* sp.
94 Red: *Herpothallon sanguineum*
95 Whitewash: *Lecidea speirea*
Dog: *Peltigera canina*
Spoon: *Cladonia gracilis*
Red-eye: *Haematomma puniceum*
Pitted: *Verrucaria calciseda*
Clot: *Mycoblastus sanguinarius*
96 Lollipop: *Pilophoron cereolus*
Map: *Rhizocarpon geographicum*
Pink: *Baeomyces roseus*
Script: *Graphis scripta*
Orange: *Caloplaca elegans*
97 Pale: *Cetraria glauca*
Boulder: *Parmelia conspersa*
Fan: *Peltigera venosa*
Shell: *Coccocarpia pellita*
Toad: *Umbilicaria papulosa*
98 Oregon: *Lobaria oregana*
Smooth: *Umbilicaria mammulata*
Horsehair: *Alectoria americana*
Golden: *Teloschistes flavicans*
Beard: *Usnea strigosa*
99 British: *Cladonia cristatella*
Pyxie: *C. pyxidata*
Spoon: *C. gracilis*
Ladder: *C. verticillata*
Reindeer: *C. rangiferina*
104 Spoon-leaved: *Sphagnum palustre*
Four-tooth: *Tetraphis pellucida*
105 Elf-cap: *Buxbaumia aphylla*
Nut Moss: *Diphyscium foliosum*
Haircap: *Polytrichum commune*
False: *Pogonatum brevicaule*
Spineleaf: *Atrichum undulatum*
106 Silky: *Dicranella heteromalla*
Burned: *Ceratodon purpureus*
Broom: *Dicranum scoparium*
Wavy: *D. rugosum*
White: *Leucobryum glaucum*
Cord: *Funaria hygrometrica*
107 Urn: *Physcomitrium turbinatum*
Apple: *Bartramia pomiformis*
Rose: *Rhodobryum roseum*
Silver: *Bryum argenteum*
Indian: *Timmia austriaca*
108 Twisted: *Tortella tortuosa*
Wall: *Tortula muralis*
White.: *Hedwigia ciliata*
Fringe: *Rhacomitrium aciculare*
109 Lum.: *Schistostega pennata*
Star: *Mnium cuspidatum*
Forest-star: *M. hornum*
Shaggy: *Rhytidiadelphus triquetrus*
110 Shiny Moss: *Heterophyllum haldanianum*
Tree: *Climacium dendroides*
Mt. Fern: *Hylocomium splendens*
Slender: *Plagiothecium denticulatum*
111 Ost.: *Ptilium crista-castrensis*
Delicate: *Thuidium delicatulum*
Feather: *Neckera pennata*
Common Water: *Fontinalis dalecarlica*
Fire-preventer: *F. antipyretica*
112 Common Hornwort: *Phaeoceros (Anthoceros) laevis*
113 Com. Liverwort: *Marchantia polymorpha*
Great: *Conocephalum conicum*
Asterella tenella
Scale: *Pallavicinia lyellii*
114 *Pellia epiphylla*
115 *Porella platyphylloidea*
Leucolejeunea clypeata
Frullania eboracensis
Scapania nemorosa
Three-lobed: *Bazzania trilobata*
Slender: *Riccia fluitans*
Purple.: *Ricciocarpus natans*
118 Fossilized leaves: *Calamites* sp.
119 Whisk: *Psilotum nudum*
Dwarf: *Equisetum scirpoides*
Giant: *E. telmateia*
120 Rough: *E. hiemale*
Smooth: *E. laevigatum*
Field: *E. arvense*

121 Wood: *E. sylvaticum*
Marsh: *E. palustre*
Water: *E. fluviatile*
122 Braun's: *Isoetes muricata*
Black-based: *I. melanopoda*
Shining: *Lycopodium lucidulum*
123 Fox-tail: *L. alopecuroides*
Stiff: *L. annotinum*
Running: *L. complanatum*
124 Wolf's Claw: *L. clavatum*
Tree: *L. obscurum*
Staghorn: *L. cernuum*
125 Rock: *Selaginella rupestris*
Underwood's: *S. underwoodii*
Resurrection: *S. lepidophylla*
126 Meadow: *S. apus*
Bigelow's: *S. bigelowii*
Oregon: *S. oregana*
127 Com. Adder's-tongue: *Ophioglossum vulgatum*
Limestone: *O. engelmanni*
128 Moonwort: *Botrychium lunaria*
Grape: *B. dissectum*
Rattlesnake: *B. virginianum*
129 Royal: *Osmunda regalis*
Cinnamon: *O. cinnamomea*
Interrupted: *O. claytoniana*
130 Hartford: *Lygodium palmatum*
Kraus': *Trichomanes krausii*
Bristle: *T. radicans*
131 Curly Grass: *Schizaea pusilla*
Coast: *Polypodium scouleri*
Licorice: *P. glycyrrhiza*
132 Strap: *Campyloneurum phyllitidis*
Res.: *Polypodium polypodioides*
Com. Polypody: *P. virginianum*
133 Leather: *Acrostichum aureum*
Golden: *Phlebodium aureum*
Vine: *Microgramma heterophylla*
134 Shoestring: *Vittaria lineata*
Ladder: *Pteris vittata*
Bracken: *Pteridium aquilinum*
135 Purple: *Pellaea atropurpurea*
Venus: *Adiantum capillus-veneris*
Maidenhair: *A. pedatum*
136 Ebony: *Asplenium platyneuron*
Maid. Spleen.: *A. trichomanes*
Wall Rue: *A. ruta-muraria*
137 Sil.: *Athyrium thelypterioides*
Lady: *A. filix-femina*
Halberd: *Tectaria heracleifolia*
138 Crested: *Dryopteris cristata*
Male: *D. filix-mas*
Marginal: *D. marginalis*
139 Amer. Tree: *Ctenitis ampla*
Marsh: *Thelypteris palustris*
New York: *T. noveboracensis*
140 Long: *T. phegopteris*
Christmas: *Polystichum acrostichoides*
Holly: *P. lonchitis*
141 Bladder: *Cystopteris fragilis*
Bulblet: *C. bulbifera*
Hay.: *Dennstaedtia punctilobula*
142 Boston: *Nephrolepis exaltata*
Virginia: *Woodwardia virginica*
Netvein: *Lorinseria areolata*
143 Hart's: *Phyllitis scolopendrium*
Ost.: *Matteuccia pensylvanica*
Oak: *Gymnocarpium dryopteris*
144 Blunt-lobed: *Woodsia obtusa*
Oregon: *W. oregana*
Woolly: *Cheilanthes tomentosa*
145 Slender: *C. feei*
Sensitive: *Onoclea sensibilis*
Walking: *Camptosorus rhizophyllus*
146 Mosquito: *Azolla caroliniana*
Floating: *Salvinia auriculata*
Water: *Marsilea quadrifolia*
147 Longleaf: *Pinus palustris*
149 Coontie: *Zamia floridana*
Sago Palm: *Cycas revoluta*
150 *Ginkgo biloba*
Green: *Ephedra viridis*
Three-leaved: *E. trifurca*
151 Atlan.: *Chamaecyparis thyoides*
Arizona: *Cupressus arizonica*
Incense: *Libocedrus decurrens*
152 Florida: *Torreya taxifolia*
Eastern Yew: *Taxus canadensis*
Monkey.: *Araucaria imbricata*
Yew: *Podocarpus macrophylla*

INDEX

MEASURING SCALE (IN MILLIMETERS AND CENTIMETERS)

MEASURING SCALE (IN 10THS OF AN INCH)
5
4
3
2
1